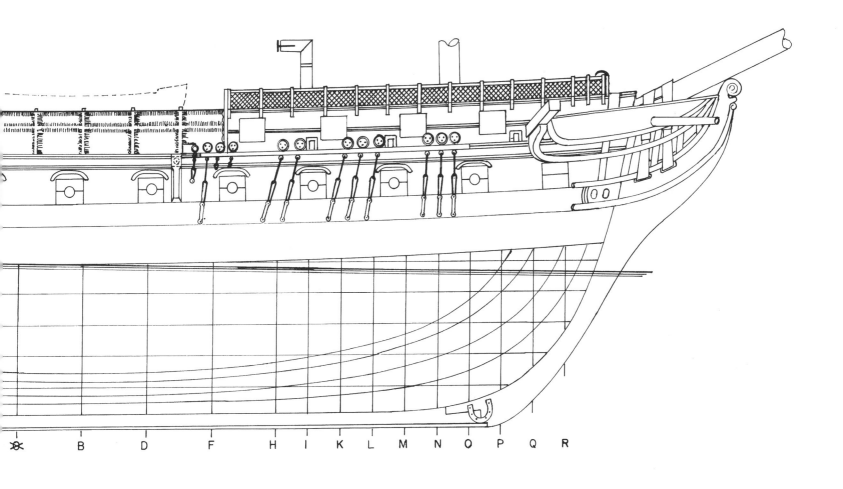

⚓   B   D   F   H   I   K   L   M   N   O   P   Q   R

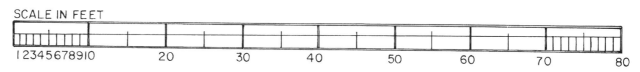

SCALE IN FEET

1 2 3 4 5 6 7 8 9 10   20   30   40   50   60   70   80

THE 44-GUN FRIGATE
USS CONSTITUTION
"Old Ironsides"

OSPREY
PUBLISHING

United States
Frigate
Constitution
44 guns

THE 44-GUN FRIGATE
USS CONSTITUTION
"Old Ironsides"

Karl Heinz Marquardt

OSPREY PUBLISHING
Bloomsbury Publishing Plc

Kemp House, Chawley Park, Cumnor Hill, Oxford OX2 9PH, UK
29 Earlsfort Terrace, Dublin 2, Ireland
1385 Broadway, 5th Floor, New York, NY 10018, USA
Email: info@ospreypublishing.com
www.ospreypublishing.com

OSPREY is a trademark of Osprey Publishing Ltd

First published in Great Britain in 2005 by Conway Maritime Press
This edition first published in Great Britain in 2017 by Osprey Publishing Ltd

A catalogue record for this book is available from the British Library

ISBN: 978 1 4728 3258 0

Editor: Nicki Marshall
Design and layout by Champion Design
Cover illustration by Ross Watton
Printed and bound in China by Toppan Leefung Printing Ltd.

23 24 25 26 27   10 9 8 7 6 5 4

Half title page
A watercolour of *Constitution* by an unknown artist, dated from about 1812. The illustration
shows the ship with a double martingale and a scroll head. (Private Boston Collection)

Frontispiece
United States 44-gun frigate *Constitution*: Sails & Rigging Plan of 1817, by Charles Ware. With
the ship laid up in ordinary at the Boston Navy Yard from late 1815 to 1821, without sails and
with most of her masts and yards lowered or removed, Ware must have relied on active frigates of
that period to construct this rig. The ensign shown is with thirteen stars and thirteen stripes, which
dates from 1777 to 1795, too early for the ship as shown. (Courtesy of the US Navy)

The Author
Karl Heinz Marquardt is an internationally acclaimed draughtsman who contributes regularly to
Conway's modelling quarterly *Model Shipwright* and has written *18th Century Rigs & Rigging* and
The Global Schooner, along with two other Anatomy of the Ship volumes on Cook's *Endeavour*
and Darwin's *Beagle*.

The Woodland Trust
Osprey Publishing supports the Woodland Trust, the UK's leading woodland conservation charity.

www.ospreypublishing.com
To find out more about our authors and books visit our website. Here you will find extracts, author
interviews, details of forthcoming events and the option to sign-up for our newsletter.

CONTENTS

FOREWORD

Putting pen to paper to tell the story and explain the lines and details of an existing ship might look like a breeze (and in some cases it is), but it becomes an obstacle course when trying to envisage only the fighting 'teenage years' of a ship that in human terms must be considered a 'Grand Old Lady of many face-lifts'. The period under observation goes back to the years of opposition between the young United States Navy and the British Royal Navy around 1812, when USS *Constitution* was a glorious war-horse. And her 1812 appearance does not match that experienced by present-day visitors aboard the surviving ship. Not only does her exterior appearance differ, but also many of the 'original pieces' – pumps, anchor-chains, capstans etc. – are from a later period of the ship's life.

Ship restoration as we know it is a twentieth-century phenomenon and unlike all those rebuilding phases the ship underwent during the first hundred years of her long life. Early major overhauls were purpose driven to update fighting capacity, to turn her later into a Navy Academy training vessel and at the end into a receiving ship. The idea of reconstructing an old pump, oven or capstan would have sounded absurd during that period; they were replaced with available new and more modern items. It was neither practicable nor fashionable for a commissioned ship to utilise a restored turn-of-the-century stern or capstan and our modern desire of reliving history just did not exist. In reality it would be much easier to reconstruct a complete 1812–1815 replica of 'Old Ironsides' from keel to truck rather than trying to re-shape her many-times-repaired surviving hull into that specific time-frame, but would it be the same?

With draughts only known from her planning stage of around 1796 and others from her major overhaul before the 1844–45 circumnavigation, today's drawings were established in 1927 and later, in short during twentieth-century reconstructions. They provide the actual lines of the ship and the status of stern and head as they appeared when the new drawings were made. Every additional part of the reconstruction of the ship in her sailing days must be considered like any other restoration: it has to be seen as an individual interpretation of known facts, whether general or specific, which means that there are many possible interpretations. In this we have to include the current sail plan, rigging arrangements, the boats and many other items.

In defence of the ship's current status as a museum ship in the Boston National Historic Park, it must be acknowledged that it is much easier to bring individual thoughts to paper than to apply them three-dimensionally to an old hull. Even with the final goal being an 1812 likeness, what can be achieved is determined by construction and financial restraints. Keeping the ship afloat has priority over any alteration of details towards a specified period. This should be understood when considering why this jewel of the US Navy is still, even after the latest reconstruction efforts, not a mirror image of her youthful years when she earned her popular nickname 'Old Ironsides'.

My attempt to re-live a certain period in the ship's extremely long life (with respect to the above notes on the interpretation of restoration plans) is based on specific documents, contemporary paintings, on general knowledge of what was possible during that era and, last but not least, on the generous help received from the Art, Curatorial and Interpretation Division of USS *Constitution*, in particular Mr William Moss, to whom I extend my heartfelt thanks. Thank you Bill!

Karl Heinz Marquardt

INTRODUCTION

'Old Ironsides' was the affectionate nickname bestowed on USS *Constitution* by her crew after a decisive thirty-minute engagement with the British 39-gun frigate HMS *Guerriere* in 1812, when they observed the enemy's round shot bouncing harmlessly off her strong oak planking without penetrating. Their shout 'Her sides must be made of iron' was soon embraced by a grateful young nation, hungry for an early victory in their new struggle with Britain.

Constitution was by then a fifteen-year-old ship, which had begun life in a time when the Continental Navy of the revolutionary years was disbanded and all ships sold to private owners. War weariness, lack of funds and fear that military forces might try to take control of the still feeble government, had this infant nation abandoning any defence force stronger then local militia. A navy, the only organisation that could have given the fledgling Union of thirteen former colonies some external cohesion and a certain standing among nations, no longer existed. The hard task of establishing the first national government began in 1789 after George Washington's inauguration as President. A national navy was not one of his early priorities, but he laid its foundations by asking the Secretary of the Treasury, Alexander Hamilton, to create a US Revenue Cutter Service of ten small vessels to reign in smuggling. Yet Congress would require much more that this to prompt them to form the nucleus of a national navy. Even when economic pressure came from belligerent nations who exploited the external weakness of the new, neutral United States and harassed the growing trade interests of their citizens, there was still no change in policy.

During the six years after the Revolution an American merchant fleet rapidly emerged, despite internal bickering and the absence of national currency or banking facilities. The fleet bartered American raw materials and some manufactured goods in harbours all over the world for merchandise needed back in the US. The European shipping establishment considered the fleet to be serious competition to their own trade, and their men-of-war found excuses to annoy and seize the US ships. Trade within the Mediterranean Sea was especially troubled by the Barbary-coast corsairs of Morocco, Algiers, Tunis and Tripoli, who preyed on helpless merchantmen not sailing under a protected flag. Unarmed ships were taken, crews captured and ransomed or sold as slaves if no payment was forthcoming. With piracy a well-established favourite pastime of these Ottoman-Empire vassals, ships of the helpless young nation from the other side of the Atlantic were welcome new quarry. After the Federal Government took office in 1789 one of their immediate concerns was the constant harassment of American shipping by North African corsairs. The matter was brought before Congress in 1790 but with little result. Two years later the problem was again raised in Congress; this time the President secured forty thousand dollars ($40,000) for ransom payments and an annual immunity payment of $25,000. However, negotiation with the corsairs broke down and acts of piracy increased. Eleven American ships were captured in the Autumn of 1793 alone and public outrage forced Congress finally into discussing the establishment of a navy. A resolution to build a few warships was passed by a very small majority, subject to the condition that in the event of a treaty with Algiers the construction of these ships should be discontinued. On 27 March 1794 the President signed and sealed the resolution to build six ships – three frigates of 44 guns and three of 36 guns; this date is today considered the birthday of the United States Navy.

The first six frigates in the US Navy:

Vessel	Size	Construction Yard
CONSTITUTION	44 guns	Boston
PRESIDENT	44 guns	New York
UNITED STATES	44 guns	Philadelphia
CHESAPEAKE	36 guns	Norfolk
CONSTELLATION	36 guns	Baltimore
CONGRESS	36 guns	Portsmouth

Work on the warships commenced, but in November 1795 a treaty with Algiers was signed. In accordance with the conditions of the treaty, $1,000,000 was paid to the Bay of Algiers for immunity against the corsairs, with further annual payments agreed. A further frigate, a brig and two schooners were promised by the US Government and later built as corsairs, and so the six warships were rendered unnecessary and left uncompleted. The President, dissatisfied with the decision to leave the frigates incomplete, declared in his 1796 Congressional Message: 'It is in our own experience that the most sincere neutrality is not sufficient guard against the depredations of nations at war. To secure respect to a neutral flag requires a naval force organised and ready to protect it from insult or aggression.'[1] After this message Congress passed an appropriation bill for the completion of the three frigates that were most advanced in construction, which became law on 20 April 1796. After all the to-ing and fro-ing of Congress, the initial three frigates were launched in 1797, the first being *United States* on 10 May at Philadelphia, followed by *Constellation* on 7 September at Baltimore, and *Constitution* on 21 October at Boston.

In the end it was not trouble with the corsairs that outraged the Nation and forced Congress to form a proper navy, but other incidents during the French–English War of 1793–1802. French privateers operated at will out of neutral American Territory, especially Charleston and Philadelphia, and captured many British and Spanish ships. In one incident alone at Charleston they destroyed a peaceful British merchantman lying at anchor and captured two American vessels. Government reaction came on 30 June 1796 through an order by the Treasury Secretary to stop privateers of all nations from entering American ports. The main reason for this order was the fear of getting embroiled in another war with Britain. France considered this a breach of their 1778 Franco–American Treaty of Alliance, believing the United States had become a British satellite. Another troubling area was French interference in the 1796 presidential election, which angered both candidates and led to an outburst by the new President John Adams. By sending emissaries to Paris, President Adams hoped to solve the French–American problems but was confronted with a demand for $250,000 in bribes, $10,000,000 as a loan and a presidential apology. Congress, although considering this an insult against the United States, were still divided on the issue, with some legislators arguing that American merchantmen were much more vulnerable during a full war and

that peace could come more easily without an official Declaration of War. So they instead opted for an undeclared quasi-war against France, which ended with cessation of hostilities between Britain and France in 1802. In this quasi-war merchantmen were not seized but sent out of US waters, yet trade with France continued without interruption.

At the outset of this quasi-war the only armed vessels the United States could muster were a few of the original ten small and inadequately armed Revenue Cutters. On 14 June 1797 Congress initially ordered the cutters to stop American citizens privateering against ships of friendly nations, thereby extending their duties to naval tasks. On 1 July, only a few weeks later, their orders were expanded to include defending the nation's coastline and repelling attacks on American ships inside territorial waters. Only two more recent additions to the small fleet, the 1797-built, 187-ton USRC *Virginia* and 98-ton USRC *General Green*, could answer this new task, with the others being either worn out or in repair. A new building programme added another eight USRC schooners by the end of the quasi-war, replacing the older vessels too worn and too small for the task. After the Congress-enforced delay in construction, the other three frigates approved in 1794 were launched about two years after the first three appeared. *Chesapeake* was launched on 20 June 1799 at Norfolk, with *President* and *Congress* on 1 April 1800 at New York and Portsmouth respectively. On 25 February 1799 the US Revenue Cutter Service (USRCS) became an integral part of the new US Navy.

By the end of the quasi-war, a new administration under President Thomas Jefferson was intent on saving money. Despite the fact that the Barbary-coast corsairs were still reaping havoc among American merchantmen, a new act was passed on 3 March 1801 on 'Providing for a naval peace establishment'. With this, all naval ships not immediately required were laid-up, repairs postponed, officers and men discharged with only skeleton maintenance crews retained. This act effectively rendered the existing navy inoperable. It forced government-owned construction facilities and navy yards to be closed or sold and brought naval ship-building to a standstill. Taking advantage of this hiatus in the navy's short history, corsair attacks on American merchantmen near Tripoli increased. In early 1803 the situation in the Mediterranean Sea came to a head and Congress authorised the construction or purchase of five vessels of war not exceeding 16 guns and the construction of fifteen gunboats. However, these government orders now had to compete with merchant ships for space in private shipyards. *Constitution* was finally repaired, after having her repairs halted in 1802 through lack of funds and being laid up for ten months, to become flagship for the Mediterranean Squadron of seven warships. The squadron continued a blockade around Tripoli in order to repel attacks on US merchant vessels and secured, in 1805, a lasting peace with the Corsair States.

When hostilities between Britain and France resumed in 1803 both combatants increased their neglect of American neutrality. In June 1807 the British frigate *Leopard* fired on the American merchantman *Chesapeake* in search of deserters, an incident that led to an embargo on all trade with Britain. President Jefferson hoped the embargo would force Britain to respect US neutrality, but in fact it devastated the nation's economy for many years to come, alienating the New England states which depended on foreign trade, and US sailors were pressed onto Royal Navy ships in increasing numbers. With national pride more offended by every incident, new president James Madison declared war on Britain on 18 June 1812, setting his much neglected small fleet against the Royal Navy, then the most powerful in the world. The ten years of disregard by an administration with little or no interest in it's navy produced a desultory fleet of nine frigates, of which two were so rotten as to not be worth repairing, three were in need of urgent repair, one was fitted out but without crew and stores (*Constitution*).and just three were commissioned. Only a handful of smaller naval vessels were at sea, mainly brigs and schooners. However, the success of single ship actions in the first few months of war developed enough national pride and confidence in the fleet to start creating a proper wartime navy.

The emergence of the US Navy can only be understood against this background of a nation struggling to find its feet in a hostile world. By the end of the quasi-war in 1801, the new US Navy comprised forty-five ships of various sizes, including eight new USRCS vessels, and had captured ninety-nine armed enemy vessels. However, it was only through the struggle with the North African Corsair states and the 1812 war with Britain that the US Navy became a major factor in the American Armed Forces.

SHIP'S HISTORY

1794

27 March George Washington signed into law an act by US Congress to authorise the building of six frigates, three of 44 guns and three of 36 guns. These ships, built to counteract Barbary Coast piracy against American merchant shipping, were the first of the emerging nation and created the fledgling Navy of the United States.

28 June Being charged with these urgent naval matters, the Secretary of War, General Henry Knox, appointed Philadelphia shipbuilder Joshua Humphreys to design the six vessels. Humphreys, a Quaker, was already active in building warships during the Revolutionary War. The contract of one of the 44-gun frigates, later named *Constitution*, was given to Edmund Hartt's shipyard in Boston, Massachusetts.

1795

Josiah Fox, a highly qualified young English shipwright with excellent connections, was appointed assistant to Joshua Humphreys. The design team also included another shipwright, William Doughty, an excellent draftsman and Humphreys' yard clerk in Philadelphia.

November With the design and preparatory work near completion, a peace treaty with Algiers stopped all work.

1796

20 April The President signed an Appropriation Bill for the completion of the three most advanced of the six frigates, one being the *Constitution* in Boston.

1797

20 September The first unsuccessful launch attempt of the 44-gun frigate USS *Constitution* was witnessed by many invited notables, including the US President. The ship moved only 27 feet down the slip and stopped. Two days later she went another 30 feet and then her stern settled down. These two futile attempts were caused by a slip that declined too moderately; declivity had to be built up to get her moving. Excessive pressure on the ship's keel during extra weeks of work on the slip brought on a permanent hog (bent keel).

21 October At the third launch attempt *Constitution* finally slid into the water. Construction space was leased at Hartt's shipyard in Boston for completion. Colonel George Claghorne was appointed naval constructor, her first captain Samuel Nicholson superintendent and General Henry Jackson naval agent. Timber used in building the ship amounted to approximately 2,000 trees, felled in forests as far apart as Maine and Georgia.

1798

27 March Outraged by the constant breach of US neutrality by French privateers, Congress voted in favour of an Appropriation Bill for fitting out the frigate as a warship. Total costs including armament were $302,719, with the annual running expenses about $125,000.

April The *Columbian Sentinel* advertised that the ship's crew would be signed on at the Federal Eagle Tavern: 'A glorious opportunity now presents to the brave and hardy Seamen of New-England to enter the service of their country – to avenge its wrongs – and to protect its rights on the ocean.' [2]

5 May Secretary of War William McHenry ordered USS Constitution to make ready for sea.

Summer The administration of all naval forces passed from the Secretary of War to the new Secretary of the Navy, Benjamin Stoddert.

| 22 July | Captain Samuel Nicholson took *Constitution* out to sea against revolutionary French privateers. Initially patrolling between Hatteras and the Florida Straits, she joined Commodore John Barry's squadron at Dominica until the end of hostilities. Here she secured West Indian waters for American shipping, thereby capturing several privateers. |

1801

Election of Thomas Jefferson, an opponent of standing armed forces, as US President. The new administration decided not to maintain the navy due to the end of the quasi-war with France and the ship was laid up at Boston. Taking advantage of this rundown of the US Navy, the Bashaw of Tripoli renewed his attacks on American merchantmen.

1802

| 6 February | Recognising that a state of war existed between the United States and Tripoli, Congress authorised the Navy to attack and seize Tripolitan shipping. Two frigates were ordered to create a blockade of Tripoli, which was relatively ineffectual. |
| 16 June | Repair on *Constitution* had to be halted due to lack of funding. |

1803

21 May	After lying for ten-and-a-half months in ordinary, the ship was repaired, fitted out and made ready for service under Commodore Edward Preble.
14 August	An order was issued for the ship to sail as flagship of the third Mediterranean squadron to the Mediterranean Sea. That squadron comprised seven ships: *Constitution* as flagship, the frigate *Philadelphia* and five smaller vessels with 16 guns or less.
12 September	The squadron arrived in Gibraltar.

8 October	Preble persuaded the departing frigates USS *New York* and USS *John Adams* to join the squadron. Together they made a show of force in Tangier Bay to remind the Sultan of Morocco, whose corsairs had become so unruly, of his obligations under the existing treaty.
24 October	*Constitution* resupplied at Cadiz and took on extra crew.
31 October	Chasing a blockade-running xebec in to inshore waters USS *Philadelphia* ran aground and was captured by nine Tripolitan gunboats, her 300 men being enslaved. A few days later the ship was towed into Tripoli harbour to be fitted out for the enemy.
29 November	*Constitution* arrived at Syracuse to establish a base for the squadron, which underwent repairs until early February. Waist planking was added to the ship, and bowsprit, boats and other parts were repaired at this time.

1804

| 16 February | Lt Stephen Decatur of USS *Constitution* sailed with sixty-five men into Tripoli harbour in a captured enemy ketch under the guidance of the Maltese pilot Salvatore Catalano. Pretending to have 'lost their anchors in a gale' they were permitted to tie up overnight alongside *Philadelphia*, and then they stormed and burnt her. Under a hail of gunfire from gunboats and the nearby fort they escaped in the small ketch without casualty. |
| 13 July | Preble's negotiations with the Neapolitan Government secured support for the US squadron. The government agreed to supply six gunboats and two bomb-vessels plus crews, a fully manned supply ship, extra armament and ammunition, plus an additional ninety-six Neapolitan sailors on temporary American payroll to improve the squadron's fighting capability. Preble departed for his blockade station off Tripoli. |

27 July	A heavy gale, which lasted several days, developed while the squadron was preparing to bombard the city. The whole fleet was battered, with some of the smaller craft near sinking.
3 August	In the afternoon the squadron arrived off Tripoli and commenced attacks on the city with USS *Constitution* taking an active part in three. In his log, Sailing Master Haraden wrote about the first day of fierce fighting:

'We were several times within two cable lengths of the rocks and within three of their batteries….two hours under fire of the enemy's batteries….we suffered most when wearing or tacking; it was then I most sensibly felt the want of another frigate….damage received in the ship is, a twenty four pound shot nearly through the centre of the mainmast, thirty feet from the deck; main royal yard and sail shot away, and our sails and running rigging considerably cut.'[3]

He also mentioned that, of the 200 shot falling within 60 feet of the ship only nine struck.

7 & 25 August	Second and third attack with *Constitution* in support but not engaging directly.
28 August	Commodore Preble described to the Secretary of the Navy one of these attacks on Tripoli by USS *Constitution* and his newly received Neapolitan gunboats:

'We continued running in, until we were within musket shot of the Crown and Mole batteries, when we brought to, and fired upwards of three hundred round shot, besides grape and canister, into the town, Bashaw's Castle, and batteries. We silenced the castle and two of the batteries for some time. At a quarter past six, the gunboats all being out of shot and in tow, I hauled off, after having been three-quarters of an hour in close action. The gunboats fired upwards of four hundred round shot, besides grape and canister, with good effect. A large Tunistan galliot was sunk in the mole – a Spanish seignior received considerable damage. The Tripolitan galleys and gunboats lost many men and were much cut.'[4]

9 September	Commodore Barron, although unwell, arrived on USS *President,* accompanied by USS *Constellation,* to relieve Commodore Preble. Preble lowered his broad pennant on *Constitution* to let *President* take over as the squadron's flagship.
12 September	At 04:30 hours, adverse winds forced *President* and *Constitution* in to collision, with the head of the latter falling on to the portside bow of the new flagship. *Constitution*'s damage was severe: flying jib- and jib booms were carried away together with the spritsail yard, the cutwater entirely broken off within the two major bobstays and the figurehead damaged beyond repair.
14 September	The ship makes for Malta to repair the damage.
17 September	Once moored in Malta's inner harbour the health boat came alongside, resulting in the yellow quarantine flag being hoisted. The crew and outside craftsmen commenced a very thorough maintenance and repair programme that lasted until 1 November.
12 October	Quarantine over.
28 October	Commodore Preble handed command of USS *Constitution* to Captain Steven Decatur and took up quarters on shore in preparation for departing home.
1 November	The squadron left Malta without Commodore Barron, whose ailing condition forced him to stay ashore to seek better medical help. Captain John Rodgers became defacto Commodore.
10 November	At Syracuse Captain John Rodgers left the smaller USS *Congress* and again made USS *Constitution* his flagship.
29 November	*Constitution* sailed alone for Lisbon to procure some good seamen, supplies and to undergo repairs.

1805

9 February — Weighed anchor to return to the Mediterranean Sea, arriving in Malta two weeks later to find Commodore Barron's health deteriorated. Supplies were replenished and the ship took over the patrol off Tripoli, reporting back to Malta on 19 March.

2 April — Departed Malta to join the full squadron. The results of a tighter Tripoli blockade became apparent in shortages of food and gun powder.

26 May — USS *Essex* joined the squadron and brought a letter from Commodore Barron handing the squadron officially over to John Rodgers. Meanwhile serious peace negotiations were under way, with a treaty prepared in Commodore Rodgers' cabin.

3 June — With the treaty signed aboard USS *Constitution*, the war against Tripoli was over and the *Philadelphia* crew released. *Constitution* sailed for Syracuse to pick up eighty-nine Tripolitan prisoners and transport them back to their home city.

27 June — *Constitution* returned to Syracuse to exchange the ship's crew (their service duty being expired) with the crew of USS *President*, which was departing for home with the gravely ill Commodore Barron on board.

7 July — Having lost a few blockade-running ships off Tripoli, the Bay of Tunis threatened the United States with war. All US ships were ordered to prepare for sea and take on several months' provisions and as much powder as the magazines could hold.

1 August — The squadron, now strongly enlarged to comprise six frigates, four brigs, two schooners, a sloop, two bomb vessels and sixteen gunboats, moved to an anchorage outside Tunis to add pressure to diplomatic talks with the Bay of Tunis.

14 August — A treaty, similar to that agreed with Tripoli, was signed by the Bay of Tunis to end attacks on American ships, and the immunity tributes, forever.

3 September — Sailing from Tunis, USS *Constitution* remained on monotonous guard duties along the western Mediterranean Sea until returning to Syracuse on 27 November.

1806

Mid-January — Following a brief trip to Malta, *Constitution* then returned to Syracuse, where she spent the next four months undergoing maintenance.

22 March — The Secretary of the Navy approved Rodgers' return home, but the President considered it important to keep USS *Constitution* with a drastically downgraded squadron in the Mediterranean Sea.

Early April — A month of lay-over at Malta before sailing to Gibraltar, arriving on 21 May.

26 May — Commodore Rodgers relinquished command of the ship, now in much need of repair, to Commodore Hugh Campbell. *Constitution* sailed first to Cadiz and then to Lisbon.

20 December — After twenty-three weeks in Lisbon she arrived at Gibraltar on her way to Syracuse via Algiers and Tunis.

1807

2 February — *Constitution* arrived back at Syracuse, leaving fourteen days later on another patrol voyage to Tunis, Algiers, Cagliari to Malta.

13 April — Remained in Malta to repair storm damage.

2 May — Departed from Malta to Syracuse where the squadron base was reduced.

| 12 June | *Constitution* left Syracuse with stops at Messina, Palermo, Leghorn, Alicante and Malaga on their slow way home. |

| 24 August | *Constitution* arrived at Algeciras, from where she departed two weeks later. |

| 8 September | Crossed the Atlantic. |

| 14 October | Arrived in Boston. |

| Mid-November | The ship sailed to New York. |

| 8 December | Command of *Constitution* was handed over to the Commandant of the New York Navy Yard, where she was partly overhauled and laid off for the next two years. |

1810

Commodore Rodgers again took command of the ship but soon transferred his flag to USS *President*, which he considered to be the faster vessel. The former first Lieutenant, Isaac Hull, became the new captain of USS *Constitution*. Only a few cruises along the American coastline were made during the year.

1811

With the new American Minister to France on board, the ship sailed towards Cherbourg where she dodged the British blockade of the port. On her return she transferred the departing minister, Mr Russel, to his new posting at the Court of St. James.

1812

Captain Hull recommended a thorough overhaul at the Washington Navy Yard, where she was careened and had her bottom cleaned and repaired. With this she became a faster ship and was well prepared for the dark clouds developing on the political horizon.

| 18 June | Flagrant illegal searches of American ships by British men-of-war increased, with members of their crews pressed into British Service. It was no longer possible for the United States to maintain strict neutrality towards both combatants, France and England; as a result war was declared against England. |

| 21 June | Although not fully equipped, *Constitution* left the Washington Navy Yard and anchored off Annapolis to take in crew and stores. They then set course towards New York to take their place in Commodore Rodgers' squadron. |

| 17 July | Sighting a small fleet near the New Jersey coast, *Constitution* tried to join them in the belief that they were part of Commodore Rodgers' squadron, but the wind died to leave them becalmed well out of range. On realising the ships to be a squadron of seven British men-of-war, Captain Hull resolved to escape or fight to the bitter end in any ensuing conflict. Despite trying everything, from wetting the sails in hope of catching even the smallest breeze to lightening the ship by dumping her drinking water, she was unable to make good an escape. While preparing to battle it out Lt Morris suggested trying to kedge her off. By extending the cables with every spare rope of the necessary diameter, they dropped the anchor by boat as far away from the ship as possible and started heaving the cable in. Noticing their enemy moving away from the fleet, the British imitated the manoeuvre but could not overtake her. On the next day when the wind reappeared *Constitution* took her advantage and escaped. |

| 19 August | A month later, and some 600 miles to the east of Boston, *Constitution* again met one of the British ships, the 39-gun frigate *Guerriere* under Captain Dacres. After an hour of manoeuvring for advantage, and occasional gunfire, the two ships came in to fighting range. Twenty minutes later the Briton's mizzenmast went over the side and the two remaining masts toppled soon after. With the ship listing and her gun-deck ports barely above the waterline, |

Captain Dacres surrendered. After the crew had been removed and taken in as prisoners, the badly damaged *Guerriere* was set alight and sank. It was during this heroic fight that USS *Constitution* acquired her nickname 'Old Ironsides': when British round shot did not penetrate her walls the sailors called out 'her sides must be made of iron'. US casualties were seven dead and seven wounded, while the British lost fifteen men, while sixty-three were wounded. In recognition of this victory Congress awarded Captain Hull a special gold medal, his officers silver medals and the crew $50,000.

15 September | After repairs at the Boston Navy Yard, Commodore Hull was relieved by Commodore William Bainbridge, the former captain of USS *Philadelphia*.

29 December | At sea and about 30 miles off the coast of Brazil, two ships were sighted, one a larger British man-of-war, the 38-gun frigate *Java*. At 14:00 hours, still a mile from the British ships, Commodore Bainbridge engaged the enemy by firing a broadside from his long guns into *Java*. Bainbridge hoped an early attack might avoid a close-quarter fight with the British carronades. Three times broadsides were fired and when *Java* made an attempt to board *Constitution* she lost her fore mast. By 16:00 hours, every mast and spar was shot away, except for the lower main mast and part of the bowsprit. After these engagements Commodore Bainbridge, who was wounded twice, drew away for some necessary repairs. On returning forty-five minutes later he found *Java* with a sail jury-rigged to her main mast's stump and as many guns as possible cleared for action. With her Captain Henry Lambert mortally wounded she surrendered within the hour. Too badly damaged to bring home as prize, *Java* had to be sunk. With his own steering wheel shot away in action Bainbridge ordered *Java*'s steering wheel to be removed and brought on board as replacement. The

casualties of this engagement were: twelve US seamen dead and twenty-two wounded; forty-eight British seamen dead and 102 wounded.

1813

End of February | Old Ironsides' victory over HMS *Java* brought great rejoicing on her return to Boston. Commodore Bainbridge and his crew received considerable recognition in the form of medals and prize money for this second triumph over the Royal Navy. Repairs were made at the US Navy Yard in Boston and, with a change of command due, Captain Charles Stewart took over while the ship was blockaded inside Boston harbour by the Royal Navy North-American and West-Indies Squadrons.

1814

January | After running the Boston blockade *Constitution* cruised under her new captain to the Windward and Leeward Islands.

14 February | Captured HM Schooner *Pictou* and three smaller vessels, returning to Portsmouth by late March.

3 April | During her voyage from Portsmouth to Boston *Constitution* found herself a few miles in lee of two British 38-gun frigates. Not to be restricted in her escape *Constitution* jettisoned extra spars, provisions and spirit casks among many other items to lighten the ship, enabling her to reach the safety of Marblehead and the protection of Fort Sewall. Returning later to Boston for repairs she was again blockaded for eight months.

December | Winter storms and the subsequent withdrawal of the British 50-gun frigate *Newcastle* from the immediate environs of Boston harbour gave Captain Stewart the chance to put to sea.

1815

20 February Standing about 180 miles off Madeira USS *Constitution* encountered HMS *Cyane*, a 22-gun sixth-rate frigate and HMS *Levant*, a 20-gun sixth-rate flush-decked ship-sloop. At sunset Captain Stewart engaged first *Cyane* and then *Levant*; the ensuing gun battle inflicted heavy damage, forcing the surrender of both ships. Taking the British officers on board *Constitution*, Captain Stewart sent prize crews and marines aboard both captured ships in order to transport them back to the United States. Casualties of this engagement were: six US dead and nine wounded; nineteen British dead and forty-two wounded.

12 March As *Constitution* lay in a light fog with her prizes at anchor off Fort Praia, Cape Verde Islands, a squadron of three larger British men-of-war, the two new 50/60-gun frigates HMS *Leander* and HMS *Newcastle* accompanied by a smaller 40-gun frigate, tried to surprise her and recapture the prizes. In a quick move Captain Stewart ordered her cables to be cut and within fifteen minutes his small fleet was on its way, separating after a short while. Giving chase in the fog, the British squadron followed *Levant* back to the Portuguese harbour and recaptured her. Thanks to the fog Old Ironsides did not end in a blaze of glory at the Cape Verde's but lived on to tell the tale.

15 May With peace declared, USS *Constitution* returned safely to New York with her prize HMS *Cyane*. The prize was purchased into the US Navy, becoming USS *Cyane*. For his victorious actions Captain Stewart was awarded a Congressional Gold Medal and the crew received a considerable amount of prize money. With this, *Constitution* became the only ship in the US Navy during the 1812 war to have Congressional Medals of valour awarded to all her wartime captains.

The names of the two Royal Navy frigates captured by *Constitution* but too damaged to bring in as prizes – *Guerriere* and *Java* – were kept alive in the US Navy, which in 1813–14 built two similarly named 44-gun frigates.

During the next six peaceful years the ship was laid up and repaired at the Boston US Navy Yard.

1821

Activated again, *Constitution* sailed under Captain Jacob Jones to the Mediterranean, taking over as flagship of the squadron stationed there. Patrol work and visits of cities filled the next three monotonous years.

1824

After being refitted during a short return to the United States and with a change of crew, Old Ironsides sailed back to the Mediterranean with Commodore Thomas Macdonough in command.

1828

19 June Decommissioned after returning to Boston Navy Yard *Constitution* was declared unseaworthy in an ensuing survey. A board of inspection recommended her to be broken up and sold. The poem 'Old Ironsides' by Oliver Wendell Holmes, published in the *Boston Advertiser* and copied across the country, stirred the public into preserving her: 'O, better than her shattered hulk should sink beneath the wave; her thunders shook the mighty deep, and there should be her grave'. Congress responded to the wave of protest and provided funds for rebuilding.

1833

24 June Under the eyes of the President and with her old captain Isaac Hull directing the ceremony, she became the first ship to enter the recently completed dry dock at Boston Navy Yard. Difficulties arose during the rebuilding phase over a contentious newly fitted figurehead of the then President, Andrew Jackson. However, the Department of the Navy did not share the objections of the City of Boston and the figurehead was installed. The rebuilding process was so thorough that it was many years before she required any major repairs.

1834

2 July   In protest over the installation of the controversial figurehead a merchant skipper, Samuel W Dewey, took advantage of the darkness one stormy night and severed the head from its torso, removing it completely. He later returned it to the Secretary of the Navy.

1835

Under Commodore Jesse D Elliot *Constitution* transported the American Ambassador to France before returning to her station as flagship of the Mediterranean squadron.

1839

After completing her tour as Mediterranean flagship, she sailed to take up the flagship position in the Pacific squadron with Commodore Alexander Claxton at the helm.

1842–43

Back on the East Coast of America she was stationed at Norfolk to serve as flagship of the Atlantic (Home) squadron.

1844–46

Following a refit at the Navy Yard at Charlestown she circumnavigated the globe under Captain John Percival. Visits were made to Havana and other ports of the West Indies. She sailed then along the coast of Brazil to Rio de Janeiro, crossing the Atlantic to Madeira, visiting the Azores, down to the Cape of Good Hope and rounding the Cape to Zanzibar and Madagascar. After leaving the Indian Ocean behind, she made further visits to Singapore, Sumatra and Borneo. The voyage continued across the Pacific to the Philippines, Canton and Hawaii. Near the end of her circumnavigation, during the Mexican War, she escorted a convoy of US merchantmen home. She was at sea for 495 days and sailed 52,370 miles.

1848

December   *Constitution*, under Captain Gwynn, again crossed the Atlantic to become flagship of the Mediterranean and African squadron. During a visit to Egypt, a son of the American consul McCauley was born aboard and named Constitution Stewart.

1849

While staying off Gaeta near Naples during a cruise along the Italian coast, Captain Gwynn extended an invitation to the King of Naples and Pope Pius IX, who fled the revolution in Rome in 1848, to visit USS *Constitution*. With the invitation accepted, the barge that brought King and Pontiff aboard was rowed by an international crew of captains from French, English, Russian, Spanish and American warships in the harbour. This was the first time that a reigning Pope visited American territory.

1851

Returning again to the United States she was laid up for the next two years in New York.

1853–55

Constitution took her position for the fifth time as flagship of the Mediterranean and African squadron. When patrolling the West African coast in search of slave traders she captured an American slave schooner. In her main duty of 'showing the flag' in many ports, she sailed 42,166 miles during her 430 days out at sea.

1855

With her operational duties coming to an end she was laid up for repairs at the US Navy Yard Portsmouth, New Hampshire.

1858

27 May   A completely overhauled *Constitution* is relaunched at the Portsmouth Navy Yard destined to serve as a training ship at the US Naval Academy in Annapolis, Maryland.

1860

1 August   Commissioned as training ship for midshipmen at Annapolis, with Lt Commander David D Porter in command.

1861

21 April   Upon outbreak of the Civil War and fearing for the ship's safety, a few companies of Massachusetts Volunteers were put on board. She ran aground while leaving the harbour on her way north to New York.

26 April   Towed into deeper water by the steamer *Boston* she began her three-day voyage in tow of the steam-gunboat *RR Cuyler*. After arrival, and for the duration of the Civil War, she resumed her duties as training ship at the resettled US Naval Academy at Newport, Rhode Island.

1865

August   With the Civil War at an end the US Naval Academy, including their training vessel *Constitution* with Commander George Dewey in charge, relocated back to Annapolis. Old Ironsides proved to be faster then her tug and was allowed to sail alone, knotting at one time 9 miles an hour; she arrived 10 hours earlier than her modern steam tug.

1871

In critical need of repair she was moved to the US Navy Yard at Philadelphia, initially to be restored for the 1876 Centennial celebrations; however, work delays, marred by shoddy workmanship, meant that she was not ready in time.

1877/78

She remained at the Philadelphia Navy Yard as a training ship.

1878

Last cruise to foreign waters. USS *Constitution* carried US exhibits for the Paris Expo to Le Havre, France. She remained for nine months at Le Havre in order to return the exhibits to the United States.

1879

16 January   On her way home from France she ran aground at Swanage Point, west of the Isle of Wight on the south coast of England. Taken off with help of her old foes, the Royal Navy, she was then docked at their facilities in Portsmouth.

24 May   *Constitution* arrived in New York, where she served for the next two years as training ship for apprentice boys, sailing in this activity to Halifax and the Caribbean. Paid off in 1881.

1881

Brought to New Hampshire, she was laid up at the US Navy Yard in Portsmouth. There she was later docked, her masts cut down and a big barn-like structure extended above the whole spar-deck. This transformed her into a giant house boat to serve as a receiving ship for new recruits.

1897

21 September   In anticipation of her 100th birthday celebrations *Constitution* was towed back to the Navy Yard at Boston, near the place she was launched and where she had received a new lease of life. The whole North Atlantic fleet

celebrated her arrival. She would not leave Boston for thirty-four years and was on exhibition for the next three.

1900

14 February Congress authorised repairs to restore the ship's hull and rigging to her former glory under the condition that the monies be donated by the public. The public response to this authorisation was only very meagre.

1905

The Navy recommended the use of the decaying hull for target practice but public outcry prevented this.

1906

Six years after Congress authorised repair, it voted $100,000 be provided for work to be done.

1907

Limited repairs were done, which included removing the accommodation structure on upper deck, and replacing a few masts and spars and much of the rigging. The ship was, however, not docked for lack of funds. She also received some replica cannon for her intended use as a national museum and remained on exhibition at the US Navy Yard Boston until 1925.

1916

Her decaying hull was by then so far deteriorated that every week 25 inches (635mm) of water had to be pumped out.

1924

Only daily pumping kept the ship afloat and experts estimated that to restore her, repairs at a cost of at least $400,000 were needed.

1925

4 March Congress authorised her restoration by public subscription. The Secretary of the Navy, Curtis Wilbur, initiated a national, voluntary campaign for restoration funds. By 1927 patriotic associations and school children raised the bulk of the $250,000 in public contributions. In response Congress voted on a donation of $920,000 for a thorough refit. For the first time a complete set of construction plans of the ship was made.

1927

16 June Within a few days of ninety-four years, and under every possible precaution, USS *Constitution*, flying from her shortened masts the flags of 1812, arrived again in the same dock she had been the first ship to enter in 1833. With only twelve per cent of her original hull timbers remaining she was then, as far as funds, materials and historical naval research made possible, restored and fitted out to her fighting-era condition. New and more accurate replica guns were also installed.

1930

15 March Refloated and with major repairs completed she left the drydock. The total costs of this first real restoration exercise were close to $1,000,000.

1931

2 July After more than three decades at the Boston Navy Yard *Constitution* departed the city on a goodwill tour to New England ports. This proved so popular that she soon left for a similar tour to all coastal states of America. Over three years, and under the command of Commander Louis G Gulliver, the ship was towed over 22,000 miles by the minesweeper USS *Grepe* and occasionally by the submarine tender USS *Bushnell* to ninety ports in twenty-one states and was visited by more than 4.6 million people.

1934

7 May *Constitution* returned home to Boston, where she still remains today, to take up her new status as 'America's Ship', representing the Nation's proud naval history.

1954

23 July President Eisenhower signed a law in which the 'Secretary of the Navy is authorised to repair, equip, and restore the United States Ship Constitution, as far as may be practicable, to her original appearance, but not for active service, and thereafter to maintain the United States Ship Constitution at Boston, Massachusetts.' [5]

1972–75

Another major restoration effort became necessary before Old Ironsides could be displayed again for the Nation's Bicentennial Celebrations in 1976. This restoration was the last performed at the Navy Yard at Boston as a working naval station. So closely connected with the ship, the Navy Yard became part of the Boston National Historic Park in 1974. The ship's restoration costs came to $4,400,000.

1976

11 July USS *Constitution* saluted the arrival of the British Royal yacht *Britannia* in Boston harbour during the Bicentennial Celebrations. This was followed by an official visit to the ship by Her Majesty Queen Elizabeth II and HRH Prince Phillip, Admiral of the Fleet, RN.

1992

25 September Again in need of a major overhaul she returned to Drydock No 1, the place she first entered 159 years earlier.

1995

26 September Three years after entering the drydock, the ship floated out of it in her best shape since her fighting days.

1997

21 July For the first time in 116 years *Constitution* sailed again. Captained by Commander Michael C Beck in this exercise outside Boston harbour, she carried six of her sails.

21 October Crew members paraded in celebration of her 200th anniversary from her birthplace, the site of the old Hartt's Shipyard, to the Old South Meeting House.

1998

21–23 July These historic dates two centuries ago, when USS *Constitution* began her active service, were commemorated with naval vessels and tall ships from around the world visiting Boston harbour. The Deputy Secretary of Defence, flying his flag aboard *Constitution*, returned with her guns the salute of visiting warships.

2000

11 July During the 'Sail Boston 2000' festivities, Old Ironsides led over 120 tall ships into Boston harbour in the Parade of Sail.

DESIGN

When President Washington took office and the first federal government was established, second to his immediate need of raising revenue came the external loss of American merchant shipping and enslavement of American sailors by the Barbary-coast corsairs. Public anger forced this to the top of the agenda and by as early as 1791 a congressional committee considered the means of setting up a naval force. However, a large number of questions – the types of ships, armament and crews required, where to build them and detailed cost estimates – remained unanswered, and the problem was handed to the Executive Branch. The Secretary of War, General Henry Knox, a Revolutionary War army officer and former bookseller with no knowledge of naval affairs, became responsible. In his discussions with advisers (friends, former naval officers, shipbuilders) it was concluded that many of the Revolutionary men-of-war were too small. In the knowledge that 44-gun ships existed in the navies of the corsair states, vessels with an equal capacity for force had to be established. With only a few ships under consideration, it was vital to make them as powerful as possible, although the economic and diplomatic pressures would limit how many large vessels there could be. The fear of American political leaders of the day, that a rise in military and naval strength could upset their still fragile hold on power, dominated even their anger over American humiliation abroad. Knox therefore proposed to Congress a new naval force with a small number of 44-gun frigates as the largest vessels. Finally a decision was reached and Congress passed an act, signed by the President on 27 March 1794, to build three frigates of 44 guns and three of 36 guns, on the condition that construction would halt when peace with Algiers, one of the Barbary-coast states, was established.

With Philadelphia being the first seat of national government, one of Secretary Knox's acquaintances was the Philadelphian shipbuilder Joshua Humphreys. Humphreys provided Knox with a half-model as part of the design proposal prior to the congressional act, and was subsequently hired to design the new frigates. The Secretary of War did not term this position as chief constructor and for reasons unknown it soon became necessary to appoint an assistant.

Joshua Humphreys, a Quaker born on 17 July 1751 in Haverford Pennsylvania, began an apprenticeship in 1765 with the known Philadelphia shipwright and builder James Penrose. In 1774 he went into partnership with his cousin John Wharton and two years later they built at their yard the *Randolph*, the only Pennsylvania-built frigate sailing under the Continental flag. She was lost on 17 March 1778 in a battle with the 64-gun HMS *Yarmouth* off Barbados with all but four hands. Humphreys is also known to have presented the Marine Committee with a number of other man-of-war draughts at this time. After the Revolutionary War he soon established himself in Philadelphia as a reputable shipbuilder and as an advocate for the country's need of a navy. He raised this subject in a number of letters before and soon after the decision was reached to build the first frigates:

'From the present appearance of affairs I believe it is time this country was possessed of a navy; but as that is yet to be raised, I have ventured a few remarks on the subject.

'Ships that compose the European navys are generally distinguished by their rates; but as the situation and depth of water of our coasts and harbors are different in some degrees from those in Europe, and as our navy for a considerable time will be inferior in numbers, we are to consider what size ship will be most formidable, and will be an overmatch for those of an enemy; such frigates as in blowing weather would be an overmatch for double-deck ships, and in light winds to evade coming to action; or double-deck ships that would be an overmatch for common double-deck ships, and in blowing weather superior to ships of three decks, or in calm weather or light winds to outsail them. Ships built on these principles will render those of an enemy in a degree useless, or require a greater number before they dare to attack our ships. Frigates, I suppose, will be the first object, and none ought to be built less than 150 feet keel, to carry twenty-eight 32-pounders or thirty 24-pounders on the gun deck, and 12-pounders on the quarter-deck. These ships should have scantlings equal to 74's, and I believe may be built of red cedar and live oak for about twenty-four pounds per ton, carpenters' tonnage, including carpenters, smiths' bill, including anchors, joiners, block makers, mast makers, riggers and rigging, sail makers and sail cloths, suits and chandlers' bill. As such ships will cost a large sum of money, they should be built of the best materials that could possibly be procured. The beams for their decks should be of the best Carolina pine, and the lower futtocks and knees, if possible of live oak.

'The greatest care should be taken in the construction of such ships, and particularly all her timbers should be framed and bolted together before they are raised. Frigates built to carry 12 and 18-pounders, in my opinion, will not answer the expectation contemplated for them; for if we should be obliged to take part in the present European war, or at a future day we should be dragged into a war with any powers of the Old Continent, especially Great Britain, they having such a number of ships of that size, that it would be an equal chance by equal combat that we lose our ships, and more particularly from the Algerians, who have ships, and some of much greater force. Several questions will arise, whether one large or two small frigates contribute most to the protection of our trade, or which will cost the least sum of money, or whether two small ones are as able to engage a double-deck ship as one large one. For my part I am decidedly of opinion the large ones will answer best.'[6]

In another letter, written after the frigates had been ordered, he wrote:

'All the maritime powers of Europe being possessed of a great number of ships of the first size contemplated, and the Algerians having several, and considering the small number of ships directed to be built, the great necessity of constructing those ships in such a way as render them less liable to be captured and more capable of rendering great services to the United States according to their number, the construction and sizes of frigates of the European nations were resorted to and their usefulness carefully considered. It was determined of importance to this country to take the lead in a class of ships not in use in Europe, which would be the only means of making our little navy of any importance. It would oblige other Powers to follow us intact, instead of our following them; considering at the same time it was not impossible we should be brought into a war with some of the European nations; and if we should be so engaged, and had ships of equal size with theirs,

for want of experience and discipline, which cannot immediately be expected, in an engagement we should not have an equal chance, and probably lose our ships. Ships of the present construction have everything in their favor; their great length gives them the advantage of sailing, which is an object of the first magnitude. They are superior to any European frigate, and if others should be in company, our frigates can always lead ahead and never be obliged to go into action, but on their own terms, except in a calm; in blowing weather our ships are capable of engaging to advantage double-deck ships. Those reasons weighed down all objections.'[7]

A young English shipwright, Josiah Fox, arrived in Philadelphia in late 1793 to study timbers and visit his extended family. From a wealthy Quaker family and born 9 October 1763 in Falmouth, England, he was apprenticed to the master shipwright of Plymouth Navy Dockyard. Of independent means, he had the opportunity after completing his apprenticeship in 1786 to study various shipyard methods, the behaviour of ships at sea and to visit foreign yards on the Continent. Twelve years younger than Humphreys and an excellent draftsman he was introduced to Knox by his cousin Andrew Ellicott, the first administrations Surveyor General. After examination by Knox and his adviser Captain John Barry, Josiah Fox was offered an opportunity to help in designing the frigates and appointed 'assistant' to Humphreys.

The third man connected with these early frigate designs was William Doughty. Yard clerk in Humphreys' shipyard and a trained shipwright and draftsman, he became Fox's assistant and was instrumental in lofting and in drawing a number of construction plans. According to Fox he also drew up one of the 36-gun frigate plans and made copies of the plans distributed to the other shipbuilders. In later years he rose to prominence and became an important name in US naval design.

Already widely discussed before design stage, the final draughts of the two classes of frigates harboured many ideas and it was up to the combined efforts of the designers to consolidate these into good fighting machines. Compared to British and French ships of similar strength they were very advanced, about 200 to 300 tons larger and instead of forecastle and quarterdeck, they were spar-deck (flush-deck) built.

It was decided that these frigates would be built not by contract to private shipbuilders but by leasing or purchasing construction places to be supervised by their captains and government-employed naval constructors. Orders for the three 44-pdrs would go to Philadelphia (Humphreys yard, J Humphreys naval constructor), to New York (leased yard, Forman Cheeseman naval constructor) and to Boston (Hartt shipyard, Col George Claghorne naval constructor). With all frigate draughts completed and issued to the respective yards construction commenced, although with some disruptions due to lack of materials, or subsequent change of specifications. The congressional peace caused a temporary stop to all building activities in the middle of construction until a supplementary act by Congress allowed the three most advanced vessels to be completed.

Criticism was raised in naval circles by several sea officers who were concerned about the ships' great size and deep draft. Certain ports, they maintained, would be too confined or shallow for the vessels, and consequently many repair facilities would become off limits. Regarding the ships' quality as fighting units, the officers also proffered their deep-rooted belief that a captain knew best how to arm his own ship. That a captain should spar and rig his ship and arm her as he saw fit was not a new idea, and was practised to a certain extent in European navies as well; however, Europe had a naval discipline established over centuries and the long learning curve of their naval officers from midshipman to captain meant that the idea was applied more efficiently than might be the case in a newly created navy. Oversparring and overgunning was a possibility with 'over-enthusiastic' officers in the young US Navy. Since a ship's performance depended as much on her trim, rig and load as on her shape, it was a captain's knowledge and experience that won the day. A good sailing ship under one command could act like a lame duck under another, and in this respect it must be said that *Constitution* was a lucky ship to have had, during her years of fighting, commanding officers of outstanding excellence.

An abundance of timber was available in the United States at the time the *Constitution*'s keel was laid; live oak came from Georgia, white oak from Massachusetts and New Jersey, yellow pine from South Carolina and red cedar, locust and white pine from Maine. Two hundred years and several restorations later, these timbers are now scarce and for many a restored part laminated timbers have to be used. Supply of good material for future restorations however is assured. At the Naval Weapons Support Centre at Crane, Indiana, the United States Navy has 30,000 acres of forest set aside for that purpose and many trees are already grown to useable size.

TABLE 1   Specifications – USS *Constitution*

Length over all	204ft 0in
Length over gundeck	174ft 10½ in
Length of keel for tonnage	145ft 0in
Moulded beam	43ft 6in
Extreme beam inside wales	44ft 2in
Depth in hold	14ft 3in
Draft	22ft 6in
Tonnage	1533^{49}/$_{94}$ tons
Thickness of oaken wooden walls	22¼ in
Armament	30 x 24-pdr long guns
	22 x 32-pdr carronades
Complement	475

TABLE 2   First complement of 400, as stated by the Navy Department

Commander	1	Gunner	1
Lieutenants	4	Quarter Gunners	11
Lieutenant of Marines	2	Coxswain	1
Sailing Masters	2	Sail maker	1
Master's Mates	2	Cooper	1
Midshipmen	7	Steward	1
Purser	1	Armourer	1
Surgeon	1	Master-at-Arms	1
Surgeon's Mates	2	Cook	1
Clerk	1	Chaplain	1
Carpenter	1	Able Seamen	120
Carpenter's Mates	2	Ordinary Seamen	150
Boatswain	1	Boys	30
Boatswain's Mates	2	Marines	50
Yeoman of gun-room	1		

TABLE 3   US 44-gun frigate of 1797 compared with a British 40/50-gun counterpart of 1795

Constitution	*Endymion*
Length over gundeck	15ft 7⅝ in shorter
Length of keel for tonnage	12ft 7½ in shorter
Breadth extreme	2ft 2⅝ in smaller
Depth in hold	1ft 11in less
Tonnage	273 tons less
Armament	only 26 x 24-pounder long guns
	same number of 32-pounder carronades
Complement 400/475 men	320 men

The concept of these US spar-deck frigates of 1797 was never fully embraced by the Royal Navy, and it was not until about fifteen years later that they were matched by new British spar-decked 60-gun 4th rates from the Leander and Newcastle class of 1813. The advantages of creating a spare-deck by joining quarterdeck and forecastle were a dry gun deck and extra room for lightweight carronades on the new deck.

DESCRIPTION

HULL

Having seen Old Ironsides in her full splendour at her current home in Boston, many visitors may ask how much of the ship is original. When considering construction materials, the answer would be a disappointing ten per cent or less; but with regard to the historic accuracy of her design, one can only answer that the finest historians in the United States have been striving for decades to achieve that goal (see Foreword). The aim is a manifestation of her appearance at the time she earned her nickname in the ongoing struggle against Britain, but the target is mighty and the road is still long. A number of elements in her exterior and interior appearance are open for debate, so an attempt will be made to explain how they fit in to the relevant years of the ship's life.

If the aim is to preserve *Constitution* in her 1812 appearance, consideration should first be given to the ship's original design and how she might have looked during her launch at Boston in 1797, rather than to any documentation of later alterations. With regard to surviving original plans, the late eminent American marine historian, Howard I Chapelle[8], spoke of a draught he believed is Humphreys' master draught for 44-gun frigates, entitled *Terrible* (one of the names he suggested for the 44s). Fox and Doughty then made copies of this with slight alterations for lofting and construction. One of the building plans has also survived[9], as has another modern draught: 'Building draught for the 44-gun frigates *Constitution, United States* and *President*' by H I Chapelle. The references supplied by Chapelle for this drawing are: '*38.4.2.A. Fox and Humphreys letters. Offsets and instructions, Doughty's copy of 1796*'.' All three draughts have only their general appearance in common. A

closer look reveals quite a few differences in the arrangement of dead eyes, the positioning of chess tree, skids and gangway steps. On the latest draught the foremost gun port or bridle port has been added to the gun deck, twelve bollard timbers are situated for and aft under the rough tree rail and a second gammoning slot is cut into the gammoning knee. It suggests that Chapelle's source, the 'Doughty copy of 1796', was based on another of the existing copies. With several copies made for the three frigates, each would have been slightly different, which is clear on the Admiralty draught of USS *President* (launched in 1800) made after capture in 1815[10], a draught completely unlike that by Chapelle. This draught is really documenting the state of the 44s during the war of 1812. Ship draughts of that era were not compatible with the ultimate drawings of today; they were only general guidelines free to individual interpretation by master shipwrights.

The appearance suggested by these earlier drawings is of a vessel with an open rough tree rail along the whole sheer from bow to stern, with fourteen or fifteen gunports laid out for long guns. *Constitution* carried twenty-eight 24-pound long guns on gundeck and ten 12-pound long guns on quarterdeck at her first appearance during the quasi-war with France. Existing pictorial and written evidence helps to establish a step-by-step change from that open rough tree rail to a boarded-up spar deck. The first change appears in 1803. The Italian ship portraitist Michele-Felice Cornè painted *Constitution* during her Boston refit for Mediterranean service, which indicates a planked-in quarterdeck topped with hammock stanchions up to the mainmast, while iron waistcloth stanchions covered the forward part of the spar deck.[11]

The second step is recorded on 19 February 1804 when she appeared at Syracuse with a planked-in waist up to the foremast shrouds.[12] The third change was the solid bulwark on the foredeck, which probably occurred after the collision with USS *President*. A note by Sailing Master Haraden stated that on 8 October 1804 (during the Malta repair period) the Carpenter substituted the guard irons with wooden rails and boarded up the space between those rails and the head.[13] However the waist planking must have been removed again during that time or soon after, since a watercolour by Commodore Rodgers from 1805 shows waistcloth between the mainmast and the foremast's aftermost shroud; all other artwork indicates that this feature remained throughout her war years. It is also visible on Charles Ware's sail plan from 1817 and in a painting of the ship's Malta visit in 1837. The first clue of a planked-up uninterrupted sheer-line, similar to the current presentation of USS *Constitution* but without an indication of spar deck gunports, is provided on the Charlestown Navy Yard draught from 1844. A photograph from 1858 prior to launching after repairs at Portsmouth indicates an appearance that had reverted back to the time before 1844. This suggests that the ship had an open, cloth-covered waist for most of her active service.

Hull dimensions can sometimes be puzzling, since there is usually a second length dimension listed on a draught. A clarification of this mysterious 'Length of keel for tonnage', is often ignored altogether in maritime books. This 'length' is in fact an artificial dimension for establishing tonnage and has nothing to do with the actual keel length. In his *Cyclopaedia* published in 1819–20, Abraham Rees clarifies the establishment of this artificial measurement:

> 'The general Rules observed for measuring the Tonnage of Ships in the Royal Navy and the Merchant's Service.
> Length of keel for tonnage is the length between perpendiculars minus 3/5 of extreme breadth, minus 2.5 inch for every foot sternpost length between upper side of keel and wing-transom.

> 'Tonnage was then calculated in the following manner:
> Length of keel for tonnage multiplied by extreme breadth, multiplied by half the extreme breadth and divided by 94.'

This method of calculating the tonnage was not very accurate and Rees suggests it was inappropriate. He concluded his explanations with: '...hence the impropriety of such a rule being made general, as it will always be found greatly to increase the tonnage of sharp-built vessels; while those that are full-built, as ships in the East India Company's service, will carry a great deal more'.[14]

QUARTER DAVITS

Quarter davits abreast the mizzenmast came into use by the end of the eighteenth century and were in general use by about 1805. The first appearance of these davits on USS *Constitution* is documented on Commodore Rodgers' watercolour of 1805. This coincides with a note that 'a gig was washed from the quarter davits February 1805'.[15] They were straight beams,

usually of pitch pine, 8 to 10 inches square and with two sheaves in their top ends. In order to lower the davit, the inner lower end was rounded, enabling it to rotate around a bolt in hinged lugs. Bringing a boat to water or taking it up again in this way was a great improvement to the older method of stowing all boats on deck. The question of why these davits were straight rather than bent like those currently on board the ship is simple. To find similarly bent timbers of equal length and strength to fit out one ship was in itself not an easy task, but to have enough spare material aboard for the carpenter to refit a broken or shot-away davit was even more difficult; straight davits were much simpler and easy to replace. A watercolour by Antoine Roux of an unknown US 36-gun frigate entering a French harbour is proof of straight davits, and we find similar evidence on the aforementioned painting of USS *Constitution* visiting the British Naval Base of Malta on 22 February 1837.[16]

FIGUREHEAD

It would be misleading to consider only one of the ship's figureheads, since in the last two centuries five different head ornaments have been used aboard *Constitution*. The first, Hercules, the symbol of strength, was carved in 1797 by the Skillings Brothers and adorned the ship's bow when she finally entered the waters of Boston harbour. Hercules was damaged beyond repair in a collision with USS *President* on 12 September 1804 during the blockade of Tripoli. It was replaced in Malta with a simple shape then generally considered to be a 'billet head'. In the 1815, enlarged, Burney edition of his *Universal Dictionary of the Marine*, William Falconer (died 1768)[17] called the shape a 'SCROWL: in ship-building, a name given to the two pieces of fir timber which are bolted to the knee of the head, and serve in lieu of a figure'. The term 'billet head' in fact encompasses two different shapes: the inward scroll like a violin was called a fiddlehead, a term F Alexander Magoun[18] used erroneously in his work for the ship's figurehead; the outward scroll, as on *Constitution*, was known as a scroll-head. When the ship arrived back in New York a figure of Neptune replaced the scroll-head, although the sea god ruled the waves only for a few years. A major overhaul at the Washington Navy Yard in 1812 brought his reign to an end and a scroll head once again adorned the ship's stem.

During *Constitution*'s 1833 reconstruction, Captain J D Elliott, the Commandant of the Boston Navy Yard, insisted on replacing the scroll head with a figurehead of Andrew Jackson, who was at that time President of the United States. As mentioned earlier, this caused great opposition in the staunchly Republican city of Boston, but with the Navy Department's approval the figurehead was carved and fitted. Many threats were made and to safeguard the figure a marine was posted nearby. However, despite all precautions, the merchant mariner Samuel W Dewey decapitated the figure and escaped undetected. Later Dewey personally handed the severed head to the Secretary of the Navy, and the damaged figure was restored to its former glory by the original carver, L S Beecher. The effigy of Andrew Jackson, with a roll of papers in his right hand and the left tucked away Napoleon style, remained in place for more than four decades until the vessel's major repairs in the 1870s at the Philadelphia Navy Yard. At this time a scroll-head was again installed, as visible in a bow view picture of 1910. Magoun mentioned that by 1927 no figurehead existed and the Navy Department decided to remain with the scroll-head, a decision adhered to by all subsequent restorations. With a scroll-head adorning USS *Constitution* for three-quarters of her long life, fitting her now with one of her former three figureheads would look rather strange.

STERN

The stern also saw alterations during *Constitution*'s long life, a fact mirrored in the stern drawings in this book. In the search for the stern of 1812 many variations emerged. Nearly all paintings of the ship, which are mostly from her glorious fighting period, show a different stern configuration. It is therefore difficult to come to an acceptable conclusion; too many variants in the number of windows (ranging from three to eight) exist, let alone taffrail shape and decorations.

In search of which artist's interpretation to trust, the foremost questions are: How many of the painters actually saw the ship and were able to make sketches? Was the painting commissioned by someone with a specific interest in the ship, or did the artist tell a story of general interest to the broader public, with accurate portrayals of the ship of secondary concern?

Paintings of Old Ironsides' battles by the Englishman Nicholas Pocock, the Frenchman Ambroise Louis Garneray or the American Thomas Birch are, for example, just retelling the story. Without knowing the extent of artistic licence used, they can not be relied upon for specific

details. However, when looking for an artist commissioned by sea officers to tell the story of their actions, one encounters the name Michele Felice Cornè. Born in Malta in1762, he was already a well-established ship portraitist when the French Revolutionary War forced him into exile. Migrating in 1800 from Italy he established himself as a painter in Salem and moved some years later to Boston. There he had several opportunities to observe *Constitution* in detail and would therefore be one of the few artists predestined to be accurate.

Cornè's first watercolour (gouache) of *Constitution* is from 1803, before she left Boston for the Mediterranean Sea. The small painting is very detailed and was probably inspired by Commodore Preble, who a few years later commissioned Cornè to paint the Battle of Tripoli. It stands to reason that such paintings for a ship's commanding officer would be as accurate as possible to satisfy a discerning customer, and to enhance the painter's reputation. The same conclusions can be drawn about another set of paintings by Cornè commemorating the action between USS *Constitution* and HMS *Guerriere*, this time commissioned by Captain Isaac Hull. Pictures by one artist but several years apart and painted under the critical gaze of the ship's captains must be considered more closely than any other battle scenes.

Although Cornè's first stern was fashioned in the style prevalent at the end of the eighteenth century, his 1812 configuration was austere by comparison. There is an unfortunate lack of detail in the two 1812 paintings in which the ship's stern is visible, and so they offer only clues to the overall concept. Cornè's earlier painting shows the stern laid out with six windows, an arrangement that was normal for ships of that size. However, his 1812 composition shows five windows, and a further work follows an identical plan. US naval historians currently claim that during the war the ship had only three windows, but to close two windows temporarily while at sea was not a hard task; it also underlines Cornè's five-window paintings. In its shape and general layout Cornè's 1812 configuration is very similar to the *President*'s stern of 1815, the only difference being six windows on the latter.

The three-window configuration currently accepted by the US Navy as the 1812 arrangement (and according to Magoun supposedly re-installed during the 1927 restoration) is visible on old photographs, for example a shot from 1903/05 with the receiving-ship house structure still above her spar deck. Considering that in 1871 *Constitution* was in critical need of repairs and paid off, laid up for ten years and then turned in to a receiving ship in 1884, this stern configuration probably goes back to repairs in the 1870s or early 1880s. That it does not go all the way back to 1812 is emphasised on the Charlestown Navy Yard draught dated July 1844. There the sheer plan specifies not the current flat stern but the more common British type still in use during the first half of the nineteenth century, with a cove above the stern windows producing a protruding taffrail. By transferring those side view heights on to a stern drawing the difference to the current outline becomes obvious. That cove above the windows is a common denominator of all contemporary *Constitution* battle-scene paintings; no matter how varied the rest of those artist's stern impressions appear.

RUDDER, PUMPS AND OTHER ITEMS

By looking over other hull features in search of the 1812 appearance, the nearest item to the stern is the rudder. Unlike the current rudder with its upper part offset and rounded, the 1812 version still had a straight and squared helm post. The rounded rudder is known as a Snodgrass rudder, named after its inventor, Gabriel Snodgrass. This invention was a big improvement over the squared helm post, altering the turning point to the centre of the rounded top part and minimising the size of the helm port. Already established for several decades on merchant ships, especially East Indiamen, it had a rather slow introduction to larger warships, not being in general use until around the mid 1820s.

The current pump system is even younger and would not have been installed before the middle of the nineteenth century. Before this date, as indicated on the Admiralty draught of the Royal Navy prize USS *President*, bilge pumping was performed by chain-pumps, and deck-wash water was lifted via elm-tree pumps from a cistern system. Sea-water tanks were located on the orlop-deck and connected via copper pipes to outboards below the waterline.

The stove was all important aboard ship. A radical break with the traditional brick fire hearth came in the 1750s, when the new and lighter iron fire hearth was introduced. In the 1780s this was replaced in the Royal Navy with the improved 'Brodie stove', which was generally still in service by 1812. It can be assumed that US frigates also used either the Brodie stove or a similar fire hearth around the same time.

ANCHORS & CAPSTAN

Arrangements for the ship's anchors regarding stowage, usage and cables around 1812 are also dissimilar to those viewed on the ship by visitors of today. Contrary to the more modern configuration where she depends on chain-cables, the ship in 1812 was still connected to her anchors via strong hemp cables. Anchor chain-cables were only tested on and introduced to British warships in 1817 and Humphreys' papers indicate that they did not appear earlier on American vessels. On a sketch by Humphreys[19] of the capstan for USS *Franklin*, launched on 21 August 1815, the lower capstan does not show any implements for using chain-cable. The invention of iron chain-cables transformed the ship's interior. They required the fitting of iron hawse-hole sleeves, of iron chain-stoppers, rounded bit-pins, a changed capstan or windlass and the creation of chain-lockers. Prior to this revolutionary change, short hemp stoppers spliced into eyebolts and lashed to the cable stopped the hemp cables from gliding out; they were belayed to riding bits and taken in with a messenger, an endless cablet of lesser diameter wound a few turns around the capstan and temporarily connected to the heavier cable. Moving down through the main hatchway onto the orlop-deck they were finally stored on cable tiers.

Although this would be accurate for 1812, to revert back to hemp cables on the museum vessel would require a massive interior reconstruction of *Constitution*. The difference in storage space alone is considerable. Chain-lockers, vertical boxed-in containments, allowed the chains to run in or out unattended; the cable tier, an orlop deck space with small timber battens nailed across, kept the wet cables off deck to let them dry out and prevent rot. Stowage of these anchor cables required the attendance of men to lay them in big loops (windings or fakes) from the outside inward so that the last part finished up in the middle, assuring an uninterrupted running out of the cable.

Anchors from that period were fitted with oaken stocks except for kedge anchors, which since the end of eighteenth century had a moveable iron stock. According to the 'Establishment and Value of Anchors for Ships of each Class in the British Navy 1809'[20], the standard number of anchors for most warships of the size of USS *Constitution* was six: sheet anchor, kedge anchor, stream anchor and three bowers. Information from the Naval Historical Centre, Boston, listed two kedge anchors but only two bowers for the ship[21], with their weight being:

Sheet anchor: 5,443.00 pd = 48.5 cwt British = 54.43 cwt American

Bowers each: 5,304.00 pd = 47.36 cwt British = 53.04 cwt American

Stream anchor: 1,100.00 pd = 9.82 cwt British = 11.0 cwt American

Kedge anchor: 700.00 pd = 6.34 cwt British = 7.0 cwt American

Kedge anchor: 403.00 pd = 3.60 cwt British = 4.03 cwt American

In total: 18,254.00 pd = 162.982 cwt (8.149 tons) British
= 182.54 cwt (9.127 tons) American

The British list of 1809[22] states for ships of 1576 tons at least: ' 4 Bowers of 52 cwt, 1 Stream anchor of 11 cwt and 1 Kedge anchor of 5.5 cwt'; the total is 11.225 tons. By comparison, 'A Table showing the Weight of Cables and Anchors, used in Ships and Vessels of War, U.S.N.'[23], gives the number of anchors for the slightly larger 44-gun USS *Brandywine* (launched 1821, 1726 tons) as 2 sheet, 2 bower and 1 stream anchor (kedge anchors not mentioned), with a total weight of 14.8 tons American = 13.214 tons British. Judging from this contemporary American and British information, the currently accepted anchor weight of USS *Constitution* seems to have been understated by approximately one third.

The anchors were stowed alongside the fore channels, with their flukes resting there and not requiring an extra fluke bed. They were kept in position by shank-painters while selvage held the oaken stock securely back in place. Cables were detached from stowed anchors and taken in.

Constitution's existing large double capstan aft of the mainmast is younger in appearance by a couple of decades and part of the anchor chain-cable transformation. The most obvious difference from a capstan used for hemp cables was the iron chain-cable transporting adaptation above the pawl rim of the lower capstan.

BOATS

When it comes to size, type and number of ship's boats, information varies greatly. Even though there was a standard set of boats for every type and rate of warship, this fluctuated with the ship's duties. A frigate serving as a squadron's flagship would have had a different assortment of boats to a normal vessel of the fleet; if she were an independent survey vessel, her complement of boats would be different again. Magoun[24] mentions up to twelve boats for frigates and described four that he believed might well have been on

Constitution when she was first commissioned: one pinnace of 24ft, a long boat of 33ft, a jolly boat of 22ft and a barge of 33ft. The most modern description of *Constitution*'s boats, found recently on the Internet, speaks of eight: a long boat of 36ft, two cutters of 30ft, two whaleboats of 28ft, a gig of 28ft, a jolly boat of 22ft and a punt of 14ft. Seven or eight is the number generally believed to be accurate. Chapelle's work[25] lists eight for *President* of 1806 and seven for a 44-gun ship of 1820:

Launch	34ft x 9ft 7in x 3ft 9in
1st cutter	32ft x 8ft x 3ft 7in
2nd cutter	28ft x 7ft x 3ft 2in
3rd cutter	26ft x 6ft 6in x 2ft 11in (stern boat)
2 quarter boats	25ft x 6ft 3in x 2ft 9in
Gig	28ft x 5ft x 2ft

The figures compiled by W Brady (in 1876) for 1st Class Frigates in a 'Table showing the Complement and Quality of Boats allowed to each Class of Vessels, U.S.N.'[26] state eight boats, with the largest a launch of 34ft. The boats could not have been larger then those listed for 44-gun ships of 1820, even on a vessel with a slightly enlarged spar-deck waist opening. It would therefore be safe to place seven boats of the listed sizes on USS *Constitution*: four in the waist, two quarter and one at the stern. In view of the frigate's flagship function, the quality of boats aboard was altered for the drawings on Ship's Boats (see pages 86–90) to a Commodore's barge of 32ft, a launch of 34ft, two cutters of 32ft and 28ft as waist boats, a 26ft cutter as stern boat, a whaleboat of 28ft and the Captain's gig of 25ft as quarter boats. The representative barge could, however, have been easily exchanged with another cutter when the flagship function ceased.

ARMAMENT

Long guns

According to Magoun[27], *Constitution*'s armament as she went in to action in the quasi-war was twenty-eight long guns of 24-pdr on gun deck and ten of 12-pdr calibre on quarterdeck. Commander Preble's journal from 14 August 1803 during the second phase of her fighting life, as flagship of the Mediterranean squadron, states: 'Sailed for Mediterranean in the Constitution

44 guns and 400 men'.[28] Further evidence is revealed in the heading of an 1803 mast and spars dimension list: 'Mounting 30. 24 Pounders & 14. 12 Pounders. As refitted in Boston – 1803'.[29] This would add six guns and bring the earlier 28/10 configuration to 30/14.

The long guns were usually cast from iron but many ships also carried a number made of brass. Brass guns were particularly employed in the binnacle area to prevent the compass needles from being influenced by the magnetic qualities of iron. *Constitution* carried a few brass guns, as indicated in various documents, including a letter from mid-1807 in which the Secretary of the Navy directed that two 24-pdr brass canons shall be transferred to USS *Wasp*.[30] The 24-pdr long guns were 9 to 9½ ft long and 47 to 50 cwt in weight, while the 12-pdr long guns were between 7 and 9ft long and between 21 and 34 cwt. They were mounted on timber carriages and required gun crews of eleven men for 24-pdr and eight for 12-pdr. One of the differences between long guns and carronades, mentioned below, was the firing range. While the former was a long-range weapon, the stubby carronade was at its best during short-range shoot-outs.

Carronades

These guns were developed in an era of close-range sea duels by combining large calibres and reduced gun weight. Cast in 1774, in the Carron Iron Work's foundry, General Melville's 'Smasher', a 68-pdr short gun of only 31 cwt (less than a 9ft 12-pdr long gun) proved such a success that it was soon produced in a range of calibres, transforming smaller vessels with normally only small-calibre weapons into formidable fighting machines. Besides weight for calibre reduction, they needed half as many crew as similar-calibre long guns. Magoun, writing mainly about the ship and her 1927 restoration, mentioned that 42-pdr carronades first replaced long guns on spar deck after the Tripoli campaign. They proved to be too heavy and when Captain Isaac Hull took command in 1810 he exchanged them with 32-pdr carronades[31]. Evidence collected by W P and E L Bass provides a better insight into this spar-deck change-over to carronades. After a failed request for carronades in the summer of 1803, Commodore Preble sent another letter to the Secretary of the Navy dated 11 March 1804[32], asking for eight 32-pdr carronades for the upper deck of *Constitution*. These arrived in late October 1804 during the repair period in Malta and were mounted in the waist. Too late for his attack

on Tripoli and lacking ammunition, they were then (on 27 November) stowed in the hold for two months before being remounted. Before returning to the United States in 1807 four of the carronades were transferred to USS *Hornet*, with the remaining four displayed in pairs on quarterdeck and forecastle. The 1812 battery of carronades comprised sixteen 32-pdr on quarterdeck and six 32-pdr in the forecastle section.

Subsequently *Constitution*'s armament changed many times until 1900 when she had none at all. A decision was made during the 1907 restoration to arm the ship with 24-pdr long guns throughout. In the next major overhaul in 1927 the spar-deck guns were replaced with 32-pdr carronades. The carronades made for this restoration were unusual, in that they were completely different from known British carronades and French obusiers and have only a slight resemblance, in their fitted trunnions and carriage, to shorter pre-1790 carronades. The shape of the newly manufactured carronades deployed on *Constitution*'s spar deck in 1927 came into vogue around 1790 but did not include the considerable changes and improvements made during that period.The post-1790 carronades were about one third longer than earlier carronades, like the length of those chosen for *Constitution* in 1927, but with a lug below the gun instead of trunnions protruding sideways, and mounted to a bed rather than a gun carriage. Another novelty of the improved carronade was the replacement, on the shorter version, of quoin and crowbar with an elevation screw, which made training for the gunner much easier and more accurate. An illustration of a carronade by Magoun[33], after an 1835 sketch by naval constructor S Hartt, shows another type again from those deployed on *Constitution* and from the type generally used in Britain. It has more in common with the experimental designs based on the outside principle of 1810 by Colonel William Congreve.[34]

Two mounting principles existed during the lifetime of carronades. Early shorter carronades followed the 'outside principle', with the undercarriage or slide being bolted or pivoted to a timber outside the gunport. The trunnions of those earlier carronades rested, like those of a long gun, on a timber carriage and were secured by capsquares. They were still trained with quoin and crowbar. The reason for using the outside principle was to bring the muzzle of these extremely short guns far enough outside of hull and rigging to prevent damage from powder burns. This arrangement required much larger gunports than normal, since carriage and large-diameter gun needed enough room for training. With the introduction, by 1790, of the improved carronade came a new mounting arrangement, the 'inboard principle'. Instead of a gun carriage it used a bed, to which the lug was bolted and a rectangular slide fitted. These early square-headed inboard slides caused many problems. As they were bolted to eyebolts below the port, traversing was not possible and metal fatigue through stress caused breakage on these eyebolts and led to carronades being overturned after firing. A Royal Navy committee was set up to solve these problems. A sketch in the committee's 1796 report on carronade mountings[35] indicated that a pivoting undercarriage with a rounded or pointed front-end would be a great improvement. It made cleaning and loading easier, training was freed from its rectangular restraints and when not in action the carronade could swing close to the sides, thereby creating more deck space.

The committee's report led to improved British carronades and French obusiers, and their findings were surely also available to the US Navy, whose captains would not have accepted inferior weapons. Since carronades were introduced to *Constitution* about ten years after these improvements had been made to British and French weapons, it seems rather strange that the carronades manufactured and installed on her in 1927 should be of a much earlier type. H I Chapelle produced a drawing of gun mountings of USS *President* in which the carronade mountings are clearly identical with the British 1796 improvements.[36] Magoun supplied dimensions for carronades of her sister ship *United States*:

Diameter of bore	6½in	Extreme diameter at breech	1ft 5¾in
Length of gun	4ft 8in	Extreme diameter at muzzle	10¼in
Length, pompellion		Length of lug	8in
To centre of lug	2ft 9in		

Further data appears in the appendix of Chapelle's later work under 'Carronade mount of the *United States* 1796':[37]

Length of skid	7ft	Aft end thick	6in
Breadth	2ft 4in	Eyebolts diameter	⅞in
Thick	7in	Diameter of eyes in bolt	3⅝in
Length of beds	4ft	Diameter in clear	1⅝in
Breadth fore end	1ft 8in	Breechings in clear	5in
Breadth aft end	1ft 10in	Bolts in skids and beds	¾in

Chapelle also lists 'Ironwork on 42-pdr carronade mount, Nov. 11, 1818' and even mentions the elevating screw as 'screw 2 $\frac{1}{2}$ in dia. ‰ 24 lbs'. Therefore the question arises: why should carronades on the spar-deck of USS *Constitution* have been different to those of other United States frigates of that period? Magoun's book[38] does not reveal the contemporary evidence that was used for reproducing these 1927 carronades. With the knowledge available in 1927, why do we have only a few 'lug-mounted' but mostly 'trunnion and sliding gun-carriage-mounted' carronades on *Constitution*'s spar-deck? This strange mix of old (odd?) and new is explained as the types used in 1805 and 1812 respectively.

RIGGING

Evidence of *Constitution*'s varying rigging throughout the first decades of her long life is provided by mast and spar dimensions of different time periods. A study of all the ship's journals would probably reveal many more details of how and when certain masts were broken or lost and at what date certain sails first appeared in her rigging. Generally, a replacement of the 'ship's engine' became necessary after several years of strenuous service. This was usually done during a major refit when masts and spars were taken off and their dimensions listed; more often than not, the dimensions of masts and spars would only be written down at the end of their working life. A new set of masts and spars did not automatically follow these old dimensions, since experience with the dispersed rig could lead to improvements. An exception to the end-of-use listing can be found in a table in Humphreys' Papers (below), which might indicate the vessel's original 1797 rig.[39] The dimensions it provides suggest that by applying British masting rules the given royal mast length can only be considered as a long pole extension to the topgallant mast. (Topgallant mast length in yards minus 3½ inch per yard divided by two-thirds = long pole length).

TABLE 4 List of Dimensions of Spars U. S. Frigate *Constitution* (Humphreys' Papers)

[Dimensions in feet]	Whole length	Mast head	Diameter at Partners	Diameter at Top floor	Diameter at Cap...
Main mast	101	15	3	2.3	1.6
Fore mast	96	14	2.10	1.11	1.6
Mizen mast	90	10	1.6		

TABLE 4 *(continued)* **List of Dimensions of Spars U. S. Frigate *Constitution***
(Humphreys' Papers)

[Dimensions in feet]	Whole length	Mast head	Diameter at Partners	Diameter at Top floor	Diameter at Cap...
Main topmast	61.6	8.9	1.7		
Fore topmast	59.6	7.9	1.7	1.2	
Mizen topmast	50.10	7	0.10	0.2	
Main topgl. mast	34.6				
Fore " "		32.9			
Mizen " "		26			
Main royal mast		18.6			
Fore " "		17			
Mizen " "		12			
Bowsprit		64 or 62 feet			
Jib boom		51	1.0		
Flying jib boom	60				
Main yard	92		1.11		
Topsail yard	64		1.5½		
Topgallant yard	44		0.11		
Royal yard	31		0.6½		
Fore yard	84		1.10		
Topsail yard	60		1.4		
Gallant yard	44.6		0.10		
Royal yard	29		0.6		
Crojack yard	64		1.5½		
Mizen tops. yard	46		0.11½		
" topgallt. yard	30		0.7		
Royal yard	20.6		0.6		
Spritsail yard	60		Slings 1.0		
Spanker boom	54		0.10		
Martingale	21.7		0.9		
Bumpkin	20		1.0		

TABLE 5 List of Dimensions of Spars U. S. Frigate *Constitution*
(Humphreys' Papers)

	Length	Width	Thickness
Main trestle trees	16	0.9½	1.7
Main topmast trestle trees	6.10	small	0.10
Crosstrees	12		
Main top	16	22	
Fore top	15	21	

TABLE 5 *(continued)*
List of Dimensions of Spars U. S. Frigate *Constitution*
(Humphreys' Papers)

	Length	Width	Thickness
Mizen top	10	15	
Fore trestle trees	15	0.9½	1.6
Fore topmast trees	6.4	0.5½	0.10½
Crosstrees	11.4		
Mizen trestle trees	10	0.7	1.1
Topmast trestle trees	4	0.6	0.8
Crosstrees			7.6
Main cap			6.8
Topmast cap	4	1.6	0.8
Fore cap	6.8		
Topmast cap	4	1.6	0.8
Mizen cap			5
Topmast cap	3	1.2	0.7

For 1803 two lists exist, one published by Howard I Chapelle with dimensions dated 1803[40] and another preserved at the National Archives, Records Group 45.[41] Both are slightly different in wording but similar in dimensions, except for fore yard (84ft to 86ft), gaff dimensions not available on the latter and the fore topmast length a writing mistake (instead of 52ft 6in, 59ft 6in). The second list, given below, is more explicit in its dimensions of trestle trees etc, more like the earliest data list in volume but slightly different in dimensions. Two main mast heights are given on this second list: one is the Humphreys dimension of 1797, which only included length, length of head and diameter at partners; the second dimension is equal to the one projected by Chapelle but included additional diameters of trestle trees and cap. The height disparity of 4ft 6in in the dimensions for the main mast probably resulted from a mast change between 1797 and 1803 when the other diameters provided on the list were lost. By trying to match other dimensions with the Humphreys list slight differences can be ascertained: the topgallant (tpg) pole in Chapelle's relatively plain list (all diameters and the extended list are missing) is still named 'Royal mast' as before, but again the length for the topgallant head is not given, nor are trestle and crosstrees mentioned for royal masts in the extended list for trees and caps, suggesting 'Royal mast' was just another name for topgallant long poles.

TABLE 6 Dimensions of the U. S. Frigate *Constitution* – Mounting 30. 24 Pounders & 14. 12 Pounders. As refitted in Boston – 1803

	Length in		Length of the Mast heads		Diameter in the Partners		Diameter in the Tresseltrees		Diameter in the Cap	
	Feet	In	Feet	In	Feet	In	Feet	In	Feet	In
Main Mast	101	—	15	—	3	—	—	—	—	—
Main Mast	105	6	15	6	3	10	2	2½	1	5 2/14
Fore Mast	96	—	14	—	2	10	1	11	1	6
Mizen Mast	90	—	10	—	1	6	—	—	—	—
Main Topmast	61	6	8	9	1	7½	—	—	—	—
Fore Topmast	59	6	7	9	1	7	1	2	—	—
Mizen Topmast	50	10	7	—	—	10	—	10	—	—
Main Top Galt. Mast	36	6	—	—	—	—	—	—	—	—
Fore Top Galt. Mast	32	9	—	—	—	—	—	—	—	
Mizen Top Galt. Mast	26	—	—	—	—	—	—	—	—	—
Main Royal Mast	18	6	—	—	—	—	—	—	—	—
Fore Royal Mast	17	—	—	—	—	—	—	—	—	—
Mizen Royal Mast	12	—	—	—	—	—	—	—	—	—
Bowsprit	64	—	2	9	—	—	—	—	—	—
Jib Boom	51	—	1	5	—	—	—	—	1	3
Flying Jib Boom	60	—	—	—	—	—	—	—	—	10

			Diameter in Slings	
Main Yard	92	—	1	11
Main Topsail Yard	64	—	1	5½
Main Top Galt. Yard	44	—	—	11
Main Royal Yard	21	—	—	6½
Fore Yard	86	—	1	10
Fore Topsail Yard	60	—	1	4
Fore Top Galt. Yard	40	6	—	10
Fore Royal Yard	29	—	—	6
Cross Jack Yard	64	—	1	5½
Mizen Topsail Yard	46	—	—	11½
Miz. Top Galt. Yard	30	—	—	7
Mizen Royal Yard	20	6	—	6
Sprit Sail Yard	60	—	1	—
Spanker Boom	50	—	—	10
Martingale	21	—	—	7 sqr.
Bumpkin	20	—	1	—
Gaff	—	—	—	—

	Length		Wide		Thick	
	Feet	In	Feet	In	Feet	In
Main Trussletrees	15	—	—	9½	1	7
Topmast ditto	6	10	11	—	10	—

TABLE 6 *(continued)*

	Length		Wide		Thick	
	Feet	In	Feet	In	Feet	In
Crosstrees	12	—	—	—	—	—
Main Top	15	—	20	—	1	3
Fore Top	15	—	20	—	1	2½
Fore Trussletrees	15	—	—	9	1	6
Topmast Trussletrees	6	4	—	10½	—	5¼
Cross Trees	11	4	—	—	—	—
Mizen Top	10	—	15—	—	11½	
Trussletrees	10	—	—	7	1	1
Topmast Trussletrees	4	—	—	6	—	8
Crosstrees	7	6	—	—	—	—
Main Cap	6	8	—	—	—	—
Top Mast Cap	4	—	1	6	—	10
Fore Cap	6	8	—	—	—	—
Top Mast Cap	4	—	1	6	—	8
Mizen Cap	5	—	—	—	—	—
Topmast Cap	3	—	1	2	—	7

The Mast & Spar Dimensions of 1803 (H I Chapelle) differs only in:

Fore Yard	84ft	0in
Fore Topmast	52ft	6in

Without any written standard rules for the masting and rigging of a ship during the early phase of the US Navy, the long-established British dimensions were relied upon, along with the modification of dimensions that might have been made by experienced captains on their own vessel. A number of examples of this practice can be found in Humphreys' Papers[42]:

Capt. Wm. Jones Proportions for Masting Frigates

Twice the beam + depth of Hold for the length of Main Mast

Main yard 8/9 of the main mast

Main topsail yard 5/7 of main yard

Main topgallant yard 2/3 of main topsail yard

Main Royal yard 1/2 of main topsail yard

Main topmast 3/5 of main mast

Main topgallant mast ⅘ of main topmast
Pole of the topgallant mast ⅖ of the whole length
The Fore masts and yards are 9⁄10 of the main masts and yards
Mizen mast 8⁄9 of the main mast
Mizen topmast ¾ of main topmast
Mizen topgallant mast ⅘ of mizen topmast
The Fore and Main mast heads ⅐ of the whole length
Main, Fore and Mizen topmast heads ⅛ of their respective lengths
Mizen mast head ⅑ of the whole length
Diameter of the masts 9⁄10 of an inch to every yard in length which is the British Rule for all Ships from 50 to 32 Guns inclusive.
Diameter of the yards 7⁄10 of an inch for every yard in length

Applying Captain Jones's rule to *Constitution* would have resulted in a main mast length of 101ft 3in with a head of approximately 14ft 6in. Besides dimensions and rules for smaller vessels and larger ships, Humphreys' Papers also contain 'Rules for Masting Frigates 1809'[43], which provide a formula for the length of main mast as '2⅕ the extreme breadth of the ship', which would bring *Constitution* to 102ft 10in. He also made a lengthy and slightly complex seven-page note of 'Commodore Rodgers Rules for Masting', dated 3 January 1815, of which the first sentence states: 'To find the length of Main Mast add twice the moulded beam to the length of Keel for Tonnage of which product take 5/11 for the length of the Mast'.[44] Commodore Rodgers' rule results in a main mast length of 105ft 6in.

In the absence of further masting directions for *Constitution* between 1803 and 1815, Chapelle's rig dimensions mirror closest those of her 'glorious years'. An early indication of this is the watercolour of the ship by Commodore Rodgers who was in command from late 1804 to 1807. Although the 1803 dimensions list a topgallant mast with long pole carrying the royal yard, Rodgers' painting of 1805 no longer shows this. The rig in the painting is already identical to Chapelle's 1815 list, with extra-stepped royal masts with long skysail-poles, and even indicates topgallant masthead lengths not mentioned in the 1815 dimensions. Painted by a seaman of high quality, the picture also indicates topgallant trestle trees and a slight offset in the mast line. A mast break with topgallant head must have existed between the topgallant mast and the royal-skysail mast, since not only would a combined and uninterrupted length of 92ft with a maximum diameter of 1ft (main topgallant mast) have been too fragile to carry three sets of sails, but also replacing a broken spar of that length at sea would have been impossible. With the royals being the highest possible sails in all listings until 1803, Rodgers placed above these royals additional triangular skysails. It is also interesting to consider Humphreys' handwritten note in the 'Rules for Masting Frigates 1809' on sky-sails and their appropriate mast lengths:

> 'Royal masts from topmast cap to hounds of top gt (gallant) mast -
> Sky scraper masts ¾ of their royal masts above royal mast head, when lowered down to step on top (gallant) mast caps.'

These rules of 1809 refer not just to sky-sail poles but also to extra stepped masts that could be lowered down. They also mention a royal mast head and a sky scraper yard of two-thirds the length of the royal yard, information which is not attainable through the contemporary listing of masts. This small note brings the date of introduction of square-rigged sky-sails forward into the end of the first decade of the nineteenth century and explains that the martingale below the bowsprit cap was to be made double, while the 1803 list provides, as expected, a single 7in square martingale (dolphin striker).

TABLE 7 As rigged until 1815 (Provided by Howard I Chapelle)

	Length	Masts Diam.	Head	Length	Yard Diam.	Arm
Fore	94ft 0in	2ft 7in	16ft 0in	81ft 0in	1ft 6in	3ft 3in
" top	56ft 0in	1ft 6½in	10ft 0in	62ft 2in	1ft ½in	5ft 3in
" topgallant	31ft 0in	0ft 11in	–	45ft 0in	0ft 9in	3ft 6in
" royal	20ft 0in	–	–	28ft 0in	0ft 7in	1ft 2in
" skysail-pole	36ft 0in	–	–	–	–	–
Main	104ft 0in	2ft 8in	19ft 6½in	95ft 0in	1ft 10½in	4ft 0in
" top	62ft 10in	1ft 6½in	10ft 0in	70ft 6in	1ft 3½in	5ft 6in
" topgallant	32ft 0in	1ft 0in	–	46ft 0in	0ft 9¾in	4ft 0in
" royal	21ft 0in	–	–	30ft 0in	0ft 8in	1ft 4in
" skysail-pole	39ft 0in	–	–	–	–	–
Mizzen	81ft 0in	2ft 1½in	13ft 6in	75ft 0in	1ft 2in	3ft 3in
" top	48ft 0in	1ft 4½in	7ft 0in	49ft 0in	0ft 9½in	4ft 0in
" topgallant	23ft 6in	0ft 9in	–	32ft 0in	0ft 7½in	2ft 6in
" royal	20ft 0in	–	–	20ft 0in	0ft 6in	3ft 0in
" skysail-pole	30ft 0in	–	–	–	–	–

TABLE 7 (continued)
As rigged until 1815 (Provided by Howard I Chapelle)

	Length	Masts Diam.	Head	Length	Yard Diam.	Arm
Spanker mast	53ft 0in	0ft 10in	–	–	–	–
Spanker boom	55ft 0in	1ft 3in	–	–	–	–
Spanker gaff	40ft 0in	1ft 2in	–	–	–	–
Bowsprit	65ft 4in	2ft 8½in	60ft 0in	1ft 2in	–	–
Jib boom	49ft 0in	1ft 2in	–	–	–	–
Flying jib boom	52ft 0in	1ft 0in	–	–	–	–

Charles Ware's 1817 sail plan of *Constitution* follows the sequence of the particular data available. There we are confronted with an even higher total mast length and with a slightly larger topgallant mast with a very long pole (drafting error). This extremely long topgallant mast and pole is shown as carrying topgallant sail, royal and square-rigged skysail plus studdingsails for topgallant and royal, which seems to be a bit much for such a relatively unsecured spire. The 'Tables showing the Masts and Spars etc. […] approved by the Secretary of the Navy 1826', probably the first official guideline for all US Navy ships, and the aforementioned 'Rules for Masting Frigates 1809' provide the evidence to declare this a draftsman's error. The list provided by these for Frigates First Class comprises: masts, topmasts, topgallant masts, royal masts, skysail masts and flagpoles. With no masthead length for the royal mast indicated, the skysail masts were actually skysail-poles (extended royal masts with flag-poles above the hounds). When comparing the lengths given for fore masts in 1815 and those in 1826, a difference of 7ft 6in is apparent, which proves that the longer mast total of Ware's sail & rigging plan is in line with the 1826 dimensions. Whereas the triangular skysails are evident in David Steel's description from 1794[46] and Commodore Rodgers' watercolour from 1805, the Ware plan indicates the square-rigged skysails mentioned in the 1809 rules and in the 1826 listing under dimensions for 'sky-sail yards'. Also newly introduced are the royal studdingsails shown on Ware's sail plan and listed by Humphreys in 'Dimensions of Spars of US Frigate *President*'[47], which includes 'Main Royal Steering Sail Boom & yard' dimensions as well as those for fore and mizzen mast. According to nautical lexica the term 'steering-sail' is an incorrect expression for studding-sail.[48]

There is a well published myth that J Humphreys gave the masts of Old Ironsides a greater diameter to make them stronger, which could only be based on two early listings. There is no evidence for this in either the 'Rules for Masting Frigates 1809'[49], which gives a diameter of masts for main and foremasts as 15/16in for every yard (1/38), or in Captain Jones' Proportions of 9/10in to 1 yard (1/40) or any other of his noted rules (1/40 as normal). Only in his 'Dimensions of Spars of US Frigate Constitution' does Humphreys list a larger (1/34) diameter (3ft in the partners for a 101ft mast). The actual dimension was 2ft 5½in, therefore 6½in less. The other 3ft diameter (listed in 1803), based on 105ft 6in in length and 2ft 8in in diameter, was four inches larger than normal, and showed an average diameter of ships above 50 guns (1/36), which *Constitution* was in size. Those dimensions were probably the reason for believing that Humphreys had stronger masts built. All contemporary authors of books on masting and rigging as well as of dictionaries, English and otherwise, agree that the largest ships had a mast diameter of 1/36 its length while frigates of 30–50 guns had a diameter of 1/40 mast length. The diameters in the 1815 table are therefore not bigger but well within the normal range for mast lengths given. Without specific US Navy ground rules on masting available for that period one would not go wrong by applying the British, especially with J Fox, one of the main contributors to these draughts, being British trained. The same can be said of Humphreys who, lacking printed material of his own, meticulously copied by hand the British Establishment's dimensions, and others, in his papers. The formula adhered to by the Royal Navy in the late eighteenth century for establishing mast height was: 'Add to the length of the gun deck the ship's extreme breadth and divide it by two. The result is the length of the mainmast.'[50]

According to this formula *Constitution*'s main mast length would have been 109ft 6in. Actual lengths of her main mast were given by Humphreys as 101ft, in the handwritten listing of 1803 as 105ft 6in, in Chapelle's 1815 list as 104ft and in the US Navy list of 1826 again as 105ft, with the 1927 listing being of similar dimensions. Therefore all were a few feet shorter in their length when compared to the Royal Navy rule of the day. By considering this and also the varying dimensions of topmasts, topgallant- and royal masts in connection with varying yard-lengths, each new rig must have had different sail shapes and sizes.

TABLES 8 Showing the MASTS AND SPARS, RIGGING AND STORES & C. of every Description, allowed to the different Classes of Vessels belonging to the NAVY OF THE UNITED STATES. Prepared by the Board of Navy Commissioners, and approved by THE SECRETARY OF THE NAVY. Washington 1826[51]

FRIGATES, First Class

Name	Length Feet	Inch	Diameter Inch	1/10	Masthead Feet	Inch
Main Mast	105	—	34	6	18	—
Top Mast	63	—	19	3	9	7
Top-gallant Mast	37	6	11	—	5	6
Royal Mast	22	—	8	8	—	—
Sky-sail Mast	37	—	6	3	—	—
Flag Pole	6	—	—	—	—	—
Fore Mast	95	—	31	5	16	—
Top Mast	56	—	19	3	9	6
Top-gallant Mast	33	6	11	—	4	6
Royal Mast	20	—	7	5	—	—
Sky-sail Mast	34	—	5		—	—
Flag Pole	6	—	—	—	—	—
Mizen Mast	84	—	24	—	12	4
Top Mast	46	4	13	5	6	8
Top-gallant Mast	24	6	8	5	4	—
Royal Mast	16	—	6	—	—	—
Sky-sail Mast	28	—	4	6	—	—
Flag Pole	5	—	—	—	—	—
Bow Sprit	66	—	—	—	—	—
Jib Boom	50	—	at cap 14	3	—	—
Flying Jib Boom	54	—	9	5	plus 4 pole	

Name	Length Feet	Inch	Diameter Inch	Yard arm 1/10	Feet	Inch
Main Yard	95	—	20	—	4	9
Top-sail Yard	71	6	16	—	6	—
Top-gallant Yard	45	—	9	5	2	6
Royal Yard	30	—	6	5	1	6
Sky-sail yard	20	—	4	5	1	—
Fore Yard	84	—	18	5	4	6
Top-sail Yard	62	—	14	7	5	3
Top-gallant Yard	41	—	9	—	2	3

Name	Length Feet	Inch	Diameter Inch	Yard arm 1/10	Feet	Inch
Fore Yard (continued)						
Royal Yard	27	—	6	—	1	4
Sky-sail Yard	18	—	4	—	—	9
Cross Jack Yard	66	—	14	—	7	—
Mizen Top-sail Yard	45	—	10	—	3	6
Top-gallant Yard	30	—	6	—	1	6
Royal Yard	19	—	4	—	—	9
Sky-sail Yard	13	—	3	—	—	6
Spritsail Yard	44	6	10	—	5	9
Spanker Boom	50	—	11	—	—	—
Mizen Gaff	32	—	8	—	plus 4 pole	
Main Gaff	30	—	8	—	—	—
Fore Gaff	36	—	8	5	—	—
Ring-tail Boom	29	2	5	8	—	—
Do. Do. Yard	14	7	3	5	—	—
Lower Swinging Boom	51	3	10	—	—	—
Do. Do. Yard	31	—	5	5	—	—
Fore Top Mast stud'g sail Boom	42	6	—	9	—	—
Do. Do. Yard	24	—	—	5	—	—
Fore Top-gallant do. Boom	32	—	6	5	—	—
Do. Do. Yard	18	—	4	—	—	—
Fore Royal do.Boom	21	—	4	—	—	—
Do. Do. Yard	12	—	3	—	—	—
Main Top Mast stud'g sail Boom	49	—	10	—	—	—
Do. Do. Yard	28	—	5	5	—	—
Main Top-gallant do.Boom	36	6	7	—	—	—
Do. Do. Yard	21	—	4	—	—	—
Main Royal do. Boom	24	6	4	5	—	—
Do. Do. Yard	14	—	3	—	—	—
Mizen Top-gallant do. Boom	22	5	4	—	—	—
Do Do. Yard	12	—	2	8	—	—
Mizen Royal do. Boom	15	—	3	—	—	—
Do. Do. Yard	8	—	2	3	—	—

35

The following list shows how the current dimensions of the ship's masts and spars compare with those of her earlier sailing years. However, the difference to earlier rigs can be seen most easily in the sail plans near the end of this book.

TABLE 9 Current list of Mast and Spars Dimensions as instigated in 1927 [52]

| Name | MASTS | | YARDS | |
	Length	Diameter	Length	Diameter
Foremast	94ft	2ft 7in	81ft	1ft 6in
Topmast	56ft	1ft 6.5in	62ft 2in	1ft 0.5in
Topgallant mast	28ft	11in	45ft	9in
Royal mast	20ft	—	28ft	7in
Mainmast	105ft	2ft 8in	95ft	1ft 10.5in
Topmast	62ft	1ft 6.5in	70ft 6in	1ft 3.5in
Topgallant mast	32ft	1ft	46ft	9.75in
Royal mast	21ft	—	30ft	8in
Mizzenmast	81ft	1ft 9.5in	75ft	1ft 2in
Topmast	48ft	1ft 2.5in	49ft	9.5in
Topgallant mast	23ft 6in	9in	32ft	7.5in
Royal mast	20ft	—	20ft	6in
Bowsprit	65ft 4in	2ft 8.25in	60ft	1ft 2in
Jib boom	49ft	1ft 2in	—	—
Flying Jib boom	52ft	1ft	—	—
Spanker mast	53ft	1ft 2in	Gaff 40ft	1ft 2in
Boom 55ft	1ft 3in			

SAILS

From all available contemporary dimension lists, the table from 1815 provides the best evidence of *Constitution*'s rig during her fighting years between 1803 and 1815. Humphreys' *Rules for Masting Frigates 1809*, which mentioned 'Skyscraper yards', is also of use. The triangular skysails in Commodore Rodgers' watercolour of 1805 could, by 1812, have already been upgraded to square sails, but the drawings given here (see pages 107, 108, 110, 124, 125) have only taken triangular skysails into account since there is no specific evidence of change by this date. These drawings reflect only on available visual and written evidence directly associated with *Constitution*. Secondary evidence from individual rules or masting lists of other ships that could have influenced the ship's rig was not taken into account. The earliest pictorial proof of skysail yards and royal studding sails on Old Ironsides comes from Ware's sail plan of 1817. It is obvious that secondary evidence would add to or alter that picture. As stated above, the *Rules for Masting Frigates 1809* in Humphreys' Papers (author unknown) reveal under 'Length of Yard: Sky Scraper 2/3 of their Royal Yard' that skysail yards came into use in or around 1809. Therefore, some ships would have carried their skysails as square sails.

Another point of interest is the 'Dimensions of Spars of the Frigate ' in Humphreys' Papers. Although no date is mentioned, the inclusion of skysail yards and their dimensions would put it very much into the same time frame. Of special interest in these dimensions for a sister-ship – USS President – is a heading 'Steering Sail Booms & Yards' (steering = studding), under which is listed 'Royal Steering Sail Boom & Yard' and this not only for main and fore mast; there is also an entry for a 'Mizen top gallant Steering Sail Boom & Yard' and a 'Mizen Royal Boom and Mizen Royal Steering Sail Yard'. Carried mainly on main and fore mast it was not unusual for studdingsails to be rigged to mizzen masts. These contemporary President spar dimensions list studdingsail booms on all topgallant yards and on the mizzen topsail yard. The dimensions of these skysail yards are, excepting miniscule differences in yard arms and diameters, within the range of the 1826 figures. By comparing President's masting list with the official 1826 US Navy table regarding studding sails, the dimensions for booms and yards are slightly different but both concur in the number of sails set.

Contrary to the current style of rigging, in which a swinging boom for a lower studdingsail is fitted to the fore channel, swinging booms in 1812 were part of the main channel. With the anchors stowed along the fore channel, the foremast's lower studdingsail was either set 'flying' with a short yard at its foot (a boat-boom could have been temporarily pushed out from the fore spar-deck to spread the foot of the studding-sail) or it was disregarded all together.

With the following detailed sails shown on the drawings in this book (see pages 120, 126, 127), the disparity between the two aforementioned lists is marked for ease of reference:

(-) not remarked on

(+) extra sails added

Head:	Fore mast:	Main fore and aft:
Flying outer jib	Fore course	Main topmast staysail
Flying jib	Fore topsail	Middle staysail
Jib	Fore topgallant sail	Main topgallant staysail
Fore topmast staysail	Fore royal sail	Main royal staysail
	Fore skysail	
	Fore lower studdingsail (-)	
	Fore top studdingsail	
	Fore topgallant studdingsail	
	Fore royal studdingsail (+)	

Main mast:	Mizzen fore and aft:	Mizzen mast:
Main course	Mizzen staysail	Mizzen topsail
Main topsail	Mizzen topmast staysail	Mizzen topgallant sail
Main topgallant sail	Mizzen topgallant staysail	Mizzen royal sail
Main royal sail	Mizzen royal staysail	Mizzen skysails
Main skysails	Spanker	
Main lower studding sail	Spanker topsail	
Main top studdingsail	Mizzen topgallant studdingsail (+)	
Main topgallant studding sail	Mizzen royal studdingsail (+)	
Main royal studdingsail (+)		

Other sails that could be added to this normal set of sails included fine weather sails, with the ringtail sail aft of the spanker and the water sail below the spanker boom.

The current rig table (of 1927) relates in many respects to that of 1815 but makes no reference to any skysail poles, in fact not even to normal flagpole extensions.

FLAGS

The modern flag of the United States is recognised all over the world, with its thirteen red-and-white alternating stripes (representing the founding states) and blue canton with 50 five-pointed stars (the current number of states). However, the composition of the flag was changed nineteen times since the first flag law of 14 June 1777. Then the Union Jack canton in the first thirteen-striped flag (Continental colours) was replaced with '13 stars white in a blue field representing a new constellation'. In 1795, after Vermont and Kentucky joined the new Republic as fourteenth and fifteenth states following the Revolutionary War, Congress decided that for each new state another star

and stripe should be added to the flag. Therefore the flag at the time of *Constitution*'s launch combined fifteen stars and fifteen stripes, and remained in this composition until 1818, when the flag reverted back to thirteen stripes with another five stars added.

As the canton arrangement of stars was not regulated until 1912, many variations appeared during the nineteenth century. Contemporary paintings of US warships from the first two decades of that century reveal a variety of flags: a four, four, three, four constellation in a slightly elongated canton; a wider canton with three rows of five stars; a square canton with five rows of three; or another square with a circle of fourteen stars around the edge and one in the centre. A few of these ensigns are also visible in paintings of *Constitution*. In lieu of a jackstaff, jacks of blue background with comparable rows or a circle of stars were hoisted either between the lower fore topmast stay and the inner jib boom or on the fore flagpole.

Besides ensign and jack, commissioned warships raised a long pennant on their main mast. There were two pennant designs. The first type of pennant comprised fifteen vertical white, red, white stripes close to the mast, with a blue swallow tail. The second design was of fifteen stars on a blue background at the mast, with a swallow tail of horizontal red and white. Flagships replaced the long pennant with a blue broad pennant, which was shorter, triangular and swallow-tailed. It carried a star design similar to the jack, with the occasional variation of fifteen stars circled around a larger Commodore star.

F A Magoun believed the navy's choice of ensign design was the five row, three star combination in a square canton: 'The ensign, known as the Fort McHenry Flag, was made under the direction of Commodore Barry and General Striker by Mrs. Mary Pickersgill, and is the navy arrangement of stars.'[53] For the ship's ensign he named a size of 16ft x 9ft 6in. The navy's broad pennant of choice was the circular one according to Magoun, of 12ft x 5ft, and the thirteen star constellation was used for the long pennant, of 40ft x 5¼6in.

This small excursion into the US naval flags of the time will bring the description of USS *Constitution* during her fighting years to an end. The description that has been gleaned from all available information may be controversial to some but hopefully informative to all. It will close with the words already stated in the Foreword: '[A reconstruction of any kind]... has to be seen as an individual interpretation of known facts, whether general or specific, which means that there are many possible interpretations.'

NOTES

1   Magoun, F A, *The Frigate Constitution and other Historic Ships* (New York, 1927) p 66

2   Heine, W C, *Historic Ships of the World* (Adelaide, S.A, 1977) p 47

3   Bass, W P & E *Constitution second phase 1802–07* (Melbourne, Florida,1982) p 40

4   Magoun, 1927, op. cit. p 70

5   Maffeo St. Cmdr. USNR, *USS Constitution Timeline* www.ussconstitution.navy (Internet, 2003)

6   Magoun, 1927, op. cit. p.64

7   Magoun, 1927, op. cit. p.65

8   Chapelle, H I, *The History of the American Sailing Navy* (New York, 1949) p 123

9   Heine, 1977, op. cit. p 46

10 Chapelle, 1949, op. cit. p 265

11 Bass, 1982, op. cit. p 20

12 Ibid. p 39

13 Ibid. p 44

14 Rees A, 'Naval Architecture 1819–20' from *The Cyclopaedia; or Universal Dictionary of Arts, Sciences and Literature, 1819–20* (reprint David & Charles Ltd, Newton Abbot, 1970) p 27

15 Bass, 1982, op. cit. p 51

16 Gruppe, H E, *The Frigates* (Time-Life Books, Amsterdam, 1979) p 165

17 Falconer, W A, *A New Universal Dictionary of the Marine, improved and enlarged by Dr. William Burney, LL.D* (London, 1815; reprint MacDonald and Jane's, London, 1974) p 436

18 Magoun, 1927, op. cit. p 89

19 Humphreys' Papers (The Historical Society of Pennsylvania) p 302

20 Falconer, repr. 1974, op. cit. p 14

21 Capstan and Anchors, www.ussconstitution.navy.mil/capstan.html (Internet, 2003)

22 Falconer, repr. 1974, op. cit. p 14

23 Brady, W N, *The Kedge Anchor or Young Sailor's Assistant*, p 492

24 Magoun, 1927, op. cit. p 103

25 Chapelle, H I, *The History of the American Sailing Navy* (New York, 1949) p 504

26 Brady, op. cit. p 513

27 Magoun, 1927, op. cit. p 101

28 Bass, 1982, op. cit. p 6

29 Ibid. p 54

30 Ibid. p 53

31 Magoun, 1927, op. cit. p 100

32 Bass, 1982, op. cit. p 53

33 Magoun, 1927, op. cit. p 100

34 Congreve, W, *An Elementary Treatise on the Mounting of Naval Ordnance* (London, 1811) p 4

35 Lavery, B, *The Arming and Fitting of English Ships of War 1600–1815* (London, 1987)

36 Chapelle, H I, *The History of American Sailing Ships* (New York, 1935) p 361

37 Chapelle, 1949, op.cit. p 516

38 Magoun, 1927, op. cit. p 100

39 Humphreys' Papers (The Historical Society of Pennsylvania) p 262

40 Chapelle, 1935, op. cit. p 484

41 Bass, 1982, op. cit. p 54

42 Humphreys' Papers (The Historical Society of Pennsylvania) p 249

43 Ibid. pp 257–259

44 Ibid. pp 278–284

45 Chapelle, 1935, op.cit. p 484

46 Gill, C S, *Steel's Elements of Mastmaking, Sailmaking and Rigging (from the 1794 edition)* (New York, 1932) p 105

47 Humphreys' Papers (The Historical Society of Pennsylvania) pp 260–261

48 Smyth, W H, *Sailor's Word Book* (London, 1867; reprint Conway Maritime Press, London, 1991) p 654

49 Humphreys' Papers (The Historical Society of Pennsylvania) pp 257–259

50 Gill, 1932, op. cit. p 45

51 *Nautical Research Journal* (Vol. 14/2/1967) p 75

52 Magoun, 1927, op. cit. p 91; and www.ussconstitution.navy.mil/spars.html (Internet, 2003)

53 Magoun, 1927, op. cit. p 102

THE PHOTOGRAPHS

Portside view of USS *Constitution* off Marblehead, Massachusetts, in 1997.

(Courtesy of the US Navy Historical Centre, Boston)

A full frontal bow view.

(Courtesy of Jim Hanna, Camberwell, Victoria)

Starboard side of the ship in Boston harbour, 1960.

(Courtesy of the US Navy Historical Centre, Boston)

Portside three-quarter bow view. (Courtesy of Jim Hanna)

A view of the portside spar-deck: in the foreground at left are the fore jeer bits and the galley funnel; situated in front of the mainmast, the tarpaulin-covered combined fore and main hatchway was used to store four of the ship's boats; in the foreground at right is the 24-pdr long chase gun mounted, with carronades behind.

(Courtesy of William Moss)

In this bow view with masts, spars and rigging, her bowsprit becomes prominent.

(Courtesy of Jim Hanna)

Three-quarter portside stern view with gallery, stern davits, main brace spread bumpkin and a bent quarter davit.

(Courtesy of William Moss, Boston)

The starboard guns: the 24-pdr long gun (chase gun) in the foreground is shown with breeching and port tackles but is not train tackle rigged; two carriage-mounted 32-pdr carronades can be seen behind.

(Courtesy of William Moss)

Crowbar and quoin-trained carriage-mounted 32-pdr carronades on starboard aft of the waist, with breeching, carriage-port and slide-port tackles also visible.

(Courtesy of William Moss)

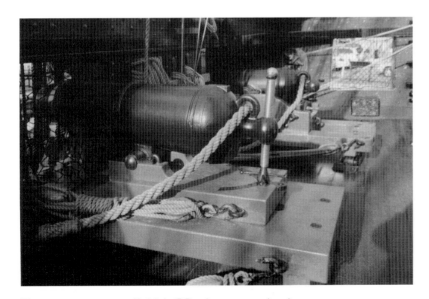

The more common British 32-pdr carronade since 1796. This carronade is lug-mounted to a bed on a pivoting slide and provides improved training with a cascable inserted training screw. The breeching and gun tackle hooked to bed and port are visible, as is the slide tackle.

(Courtesy of William Moss)

41

The galley oven on gun-deck, which is situated behind the fore mast. Chain-cable bits can be seen in the foreground to both sides. (Courtesy of William Moss)

Two views of the steering wheel of *Constitution* while undergoing repairs in the workshop.

(Courtesy of William Moss)

A 24-pdr long gun on the gun-deck, with breeching, gun tackle and quoin. The inboard running train tackle is not fitted for public safety. A water bucket is situated between the guns to wet the sponge for cleaning the bore from gun powder residues before reloading. (Courtesy of William Moss)

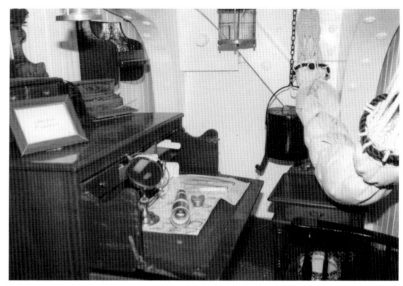

The Fifth Lieutenant's cabin.

(Courtesy of William Moss)

The lower capstan on gun-deck aft of the mainmast. The blackened lower section comprises the pawl rim and a chain-cable transporting adaptation above it.

(Courtesy of William Moss)

The ward room, with panels and doors at both sides leading to the officers cabins, the captain's pantry aft of the mizzen mast and the bulkhead and door to the captain's quarters.

(Courtesy of William Moss)

Captain's day cabin: the ward room bulkhead and a
24-pdr long gun are situated at left; two gun ports are
on the starboard side, one with a carriage-mounted gun.

(Courtesy of William Moss)

A view of the surgeon's cockpit.

(Courtesy of William Moss)

Two hammocks slung to the deck beams
on berth deck, the living quarters of the crew.

(Courtesy of William Moss)

After gun powder filling room. The rack contains filled
and empty cartridge bags, slippers, copper powder ladle
and a small powder keg.

(Courtesy of William Moss)

Old Ironsides on a slip a few days before re-launching on 27 May 1858 after a major repair at Portsmouth, New Hampshire. The ship was being prepared for service as training ship at the US Naval Academy, Annapolis; her copper sheathing is, as this stage, still incomplete.

(Courtesy of the US Navy)

The ship at the Boston Navy Yard pier after her first reconstruction and before her good-will tour in 1931.

(Courtesy of the US Navy Historical Centre, Boston)

Constitution with her receiving ship superstructure in 1905 in Boston harbour.

(Courtesy of the US Navy)

A watercolour of *Constitution* painted by her captain, Commodore John Rodgers, between November 1804 and May 1806.

(Courtesy of the US National Archives)

A General Arrangement

A1 RECONSTRUCTION OF 1812

A1/1 Sheer elevation

A1/2 Half breadth plan

A1/3 Body plan

A1/1

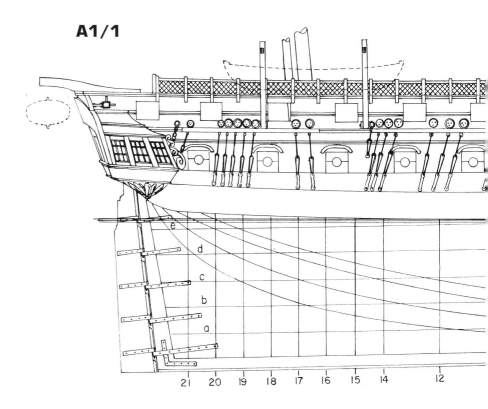

A1/3

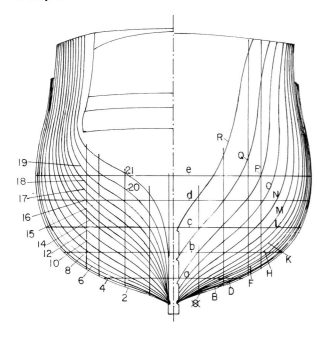

A1/2

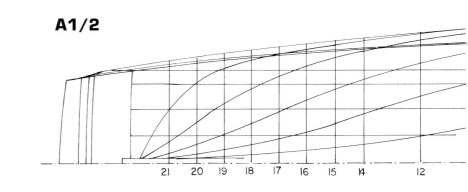

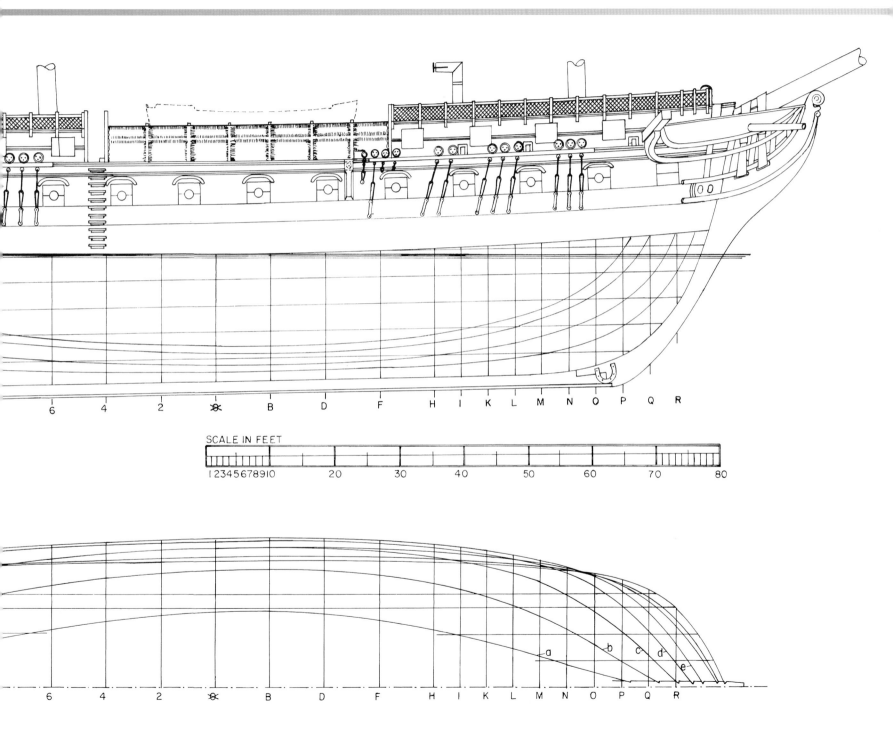

SCALE IN FEET

1 2 3 4 5 6 7 8 9 10   20   30   40   50   60   70   80

A General Arrangement

A1/4 PROFILE

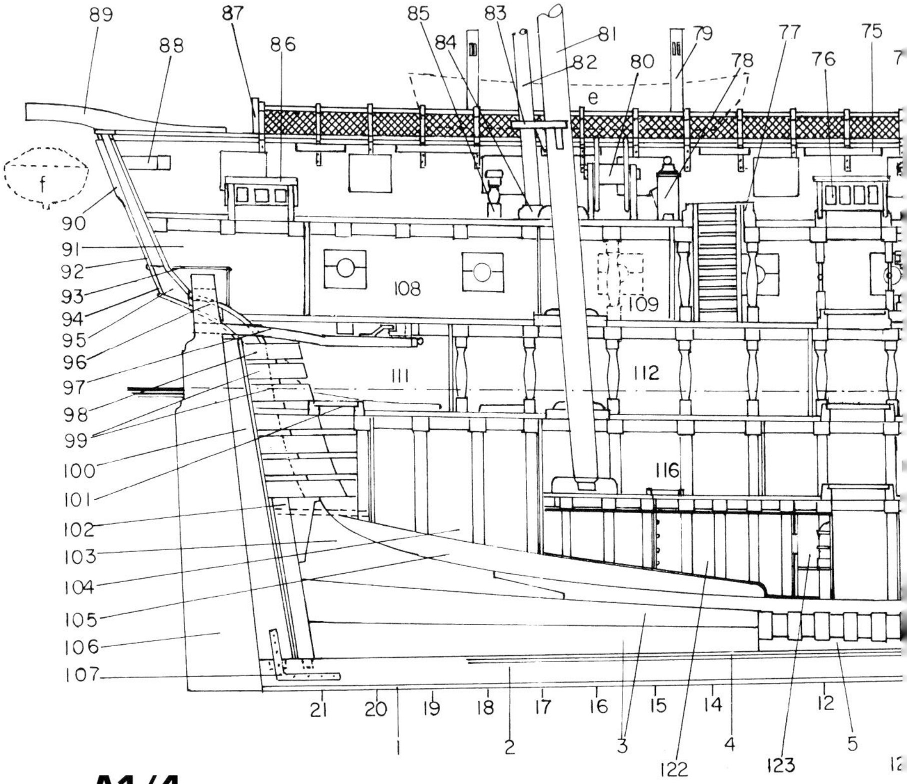

1 False keel
2 Keel
3 Aft deadwood
4 Rabbet
5 Rising wood
6 Frames
7 Keelson
8 Mainmast step
9 Fore deadwood
10 Forefoot
11 Stem
12 Horseshoe iron
13 Gripe
14 Foremast step
15 Breast hook
16 Apron
17 Stemson
18 Upper main piece
19 Chocks
20 Main piece lacing
21 Gammoning knee
22 Lower part of bobstay piece
23 Upper part of bobstay piece
24 Gammoning holes
25 Scroll head
26 Bowsprit
27 False rail
28 Cross-pieces and head grating
29 Gunwale
30 End-board and opening to head
31 Pin rail
32 Manger
33 Cover-board
34 Fore peak
35 Cathead
36 Bowsprit step
37 Scuttle
38 Fore topsail sheet bits
39 Foremast

40 Hammock netting
41 Fore jeer bits
42 Second riding bits
43 Top timber opening for anchor lashing
44 Pin rail
45 Hammock rail stanchion
46 Caboose grating
47 Fire place
48 First riding bits
49 Carronade port
50 Head ledge on spar-deck opening
51 End-board for hammock rails
52 Waist rail ropes
53 Coaming on spar-deck opening
54 Stairway to gun-deck
55 Fore hatchway
56 Boat chocks
57 Gun port
58 Deck beam
59 Gun-deck beam stanchion
60 Waist rail stanchion
61 Main hatchway
62 Crank-handle for chain pump
63 Stairway to gun-deck
64 Centre gang-plank
65 Fore chain pump
66 Gangway end-boards
67 Main topsail sheet bits
68 Mainmast
69 Main jeer bits
70 Twin chain pumps
71 Spar-deck
72 Aft stair- hatchway
73 Removable stanchions
74 Capstan
75 Pin rail
76 Hatchway with skylight to wardroom

A1/4

77 Companion
78 Binnacle
79 Quarter boat davits
80 Steering wheel
81 Mizzen mast
82 Spanker mast
83 Spanker boom crutch
84 Spanker mast step
85 Mizzen jeer bits
86 Cabin skylight
87 Hammock cradle end-board

88 Taffrail support knee
89 Stern davits
90 Taffrail
91 Captain's inner cabin
92 Stern windows
93 Rudder head cover
94 Upper counter
95 Stern timbers
96 Rudder head in helm-port
97 Tiller
98 Wing transom
99 Transoms

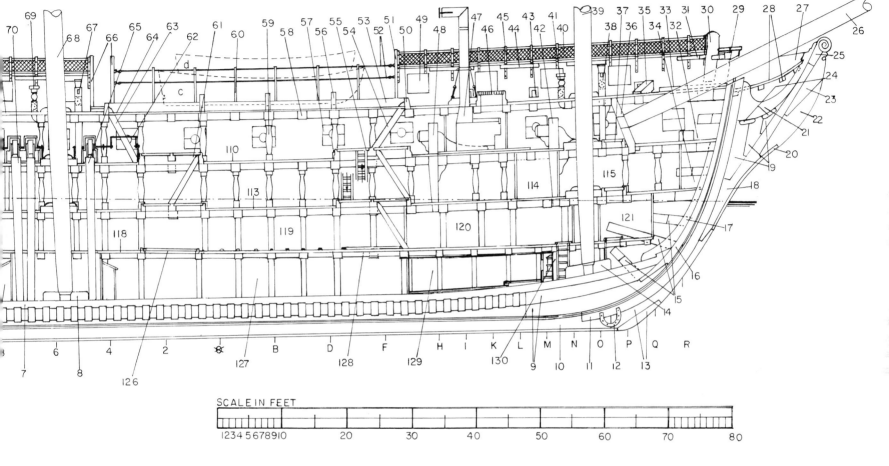

SCALE IN FEET

1 2 3 4 5 6 7 8 9 10   20   30   40   50   60   70   80

100 Sternpost
101 Scuttle to lady's hole
102 Inner sternpost
103 Sternson
104 Bread room
105 Deadwood
106 Rudder
107 Iron heel bands
108 Captain's day cabin
109 Captain's pantry
110 Gun-deck
111 Officers pantry and storage

112 Wardroom
113 Berth-deck
114 Brick
115 Hospital
116 Filling room
117 Cockpit
118 Orlop-deck
119 Cable tier
120 Sail room
121 Gunner's room
122 Aft powder room
123 Aft light room

124 Spirit room
125 Shot locker
126 Lower main hatchway
127 General storage hold
128 Lower fore hatchway
129 Forward powder room
130 Forward light room

A General Arrangement

A1/5 Spar-deck

1 Stern davit
2 Rough tree rail
3 Taffrail support knee
4 Main brace spreader
5 Gallery
6 Gunwale
7 Cabin skylight
8 Hammock rail
9 Mizzen channel
10 Mizzen jeer bits
11 Spanker mast step
12 Spanker mast
13 Mizzen mast
14 Steering wheel
15 Binnacle
16 Quarter davits
17 Companion
18 Pin rail
19 Carronade location
20 Main channel
21 Wardroom skylight
22 Capstan
23 Aft stair-hatchway
24 Main jeer bits
25 Mainmast
26 Main topsail sheet bits
27 Gangway end-boards
28 Gangway
29 Aft ledge of spar-deck
   opening
30 Aft stairway to gun-deck
31 Centre gang-plank
32 Deck beam
33 Side coaming of spar-deck
   opening

34 Boat chocks
35 Waist stanchion
36 Waist cloth
37 Waist stanchion ropes
38 Spar-deck opening
39 Chess-tree
40 Fore stairway to gun-deck
41 End-board of hammock cradle
42 Fore channel
43 Fireplace funnel
44 Caboose grating
45 Hammock rail
46 Fore jeer bits
47 Foremast
48 Chase gun location
49 Fore topsail sheet bits
50 Cathead
51 Cover-board
52 Pin rail
53 Fore end-board
54 Entrance to head
55 Boomkins
56 Seats of ease
57 False rail
58 Head
59 Bowsprit

Boat location

a 34ft launch
b 32ft Commodore's barge
c 32ft cutter
d 28ft cutter
e 26ft cutter
f 26ft Captain's gig
g 28ft whale boat

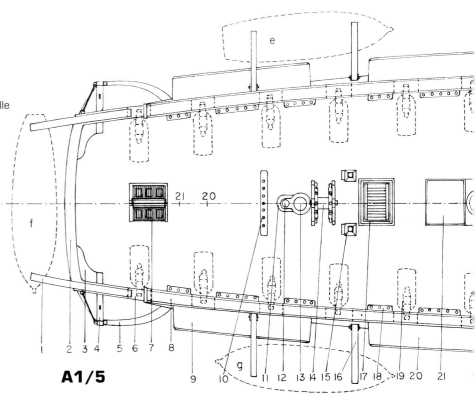

A1/5

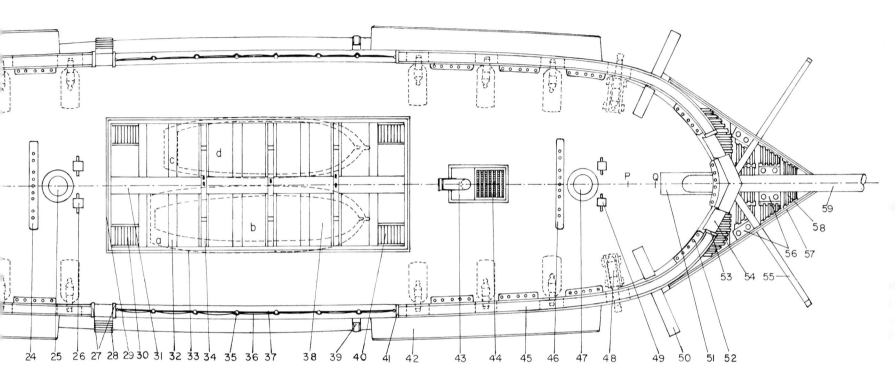

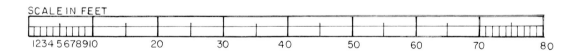

24　25　26　27　28　29　30　31　32　33　34　35　36　37　　38　39　40　41　42　　43　　44　　45　　46　　47　　48　　49　50　51　52

SCALE IN FEET

1 2 3 4 5 6 7 8 9 10　　　20　　　30　　　40　　　50　　　60　　　70　　　80

A General Arrangement

A2 ORIGINAL 1797
DRAUGHTS

A2/1 Sheer elevation of H I
Chapelle's interpretation
of this draught

A2/2 Sheer elevation of
original 1795 draught
(Doughty copy)

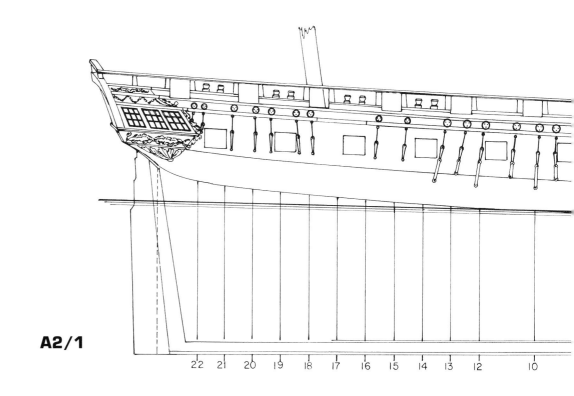

A2/1

22 21 20 19 18 17 16 15 14 13 12 10

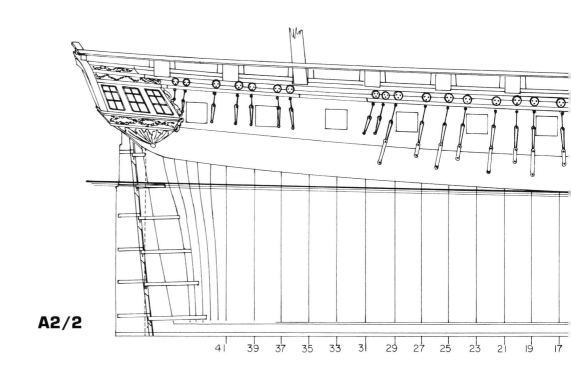

A2/2

41 39 37 35 33 31 29 27 25 23 21 19 17

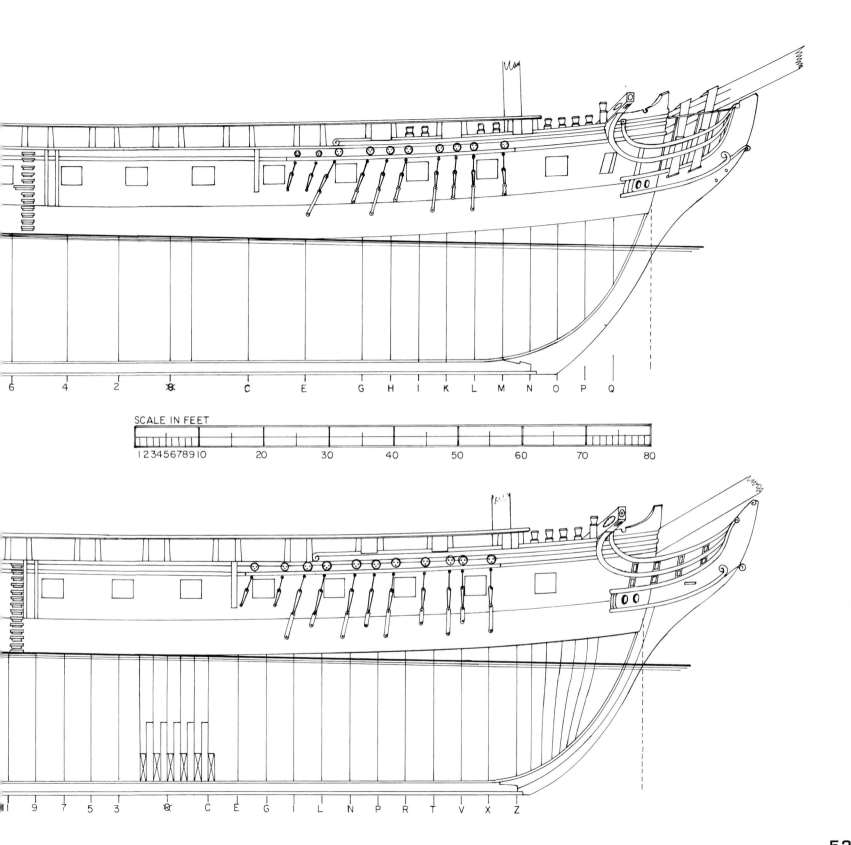

6 4 2 ⊗ C E G H I K L M N O P Q

SCALE IN FEET

1 2 3 4 5 6 7 8 9 10   20   30   40   50   60   70   80

1 9 7 5 3 ⊗ C E G I L N P R T V X Z

A   General Arrangement

A2/3 Vertical lines drawing

A2/4 Half breadth lines

A2/5 Body lines

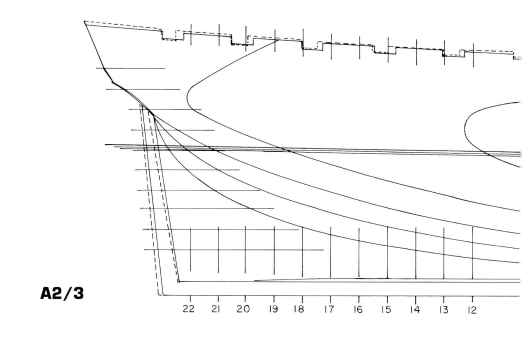

A2/3

A2/5

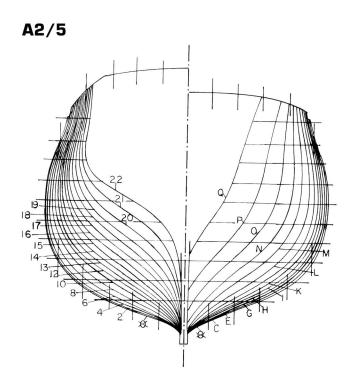

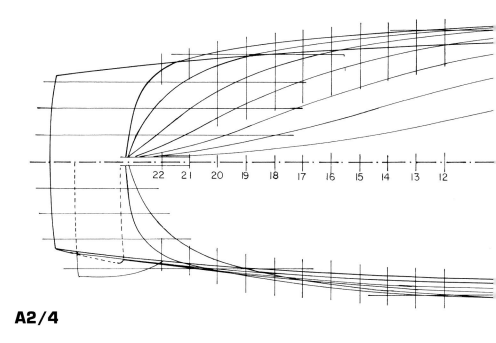

A2/4

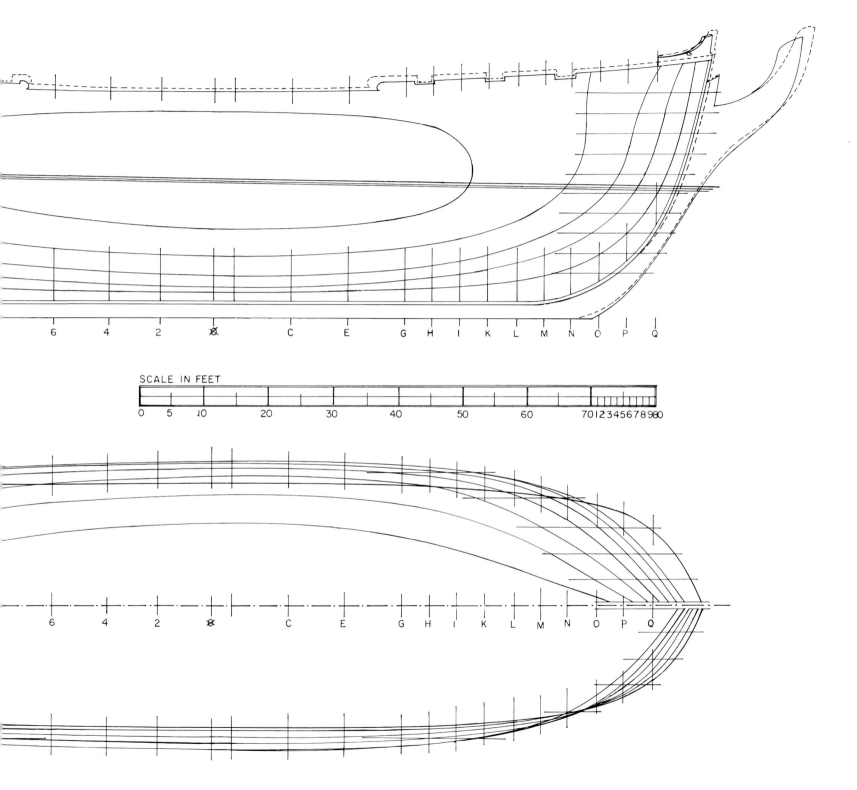

6   4   2   ⊗   C   E   G H I K L M N O P Q

SCALE IN FEET

0   5   10   20   30   40   50   60   7012345678980

6   4   2   ⊗   C   E   G H I K L M N O P Q

55

A   General Arrangement

A3   TWENTIETH-CENTURY
RESTORATION DRAWINGS

A3/1   Sheer elevation

A3/2   Vertical lines drawing

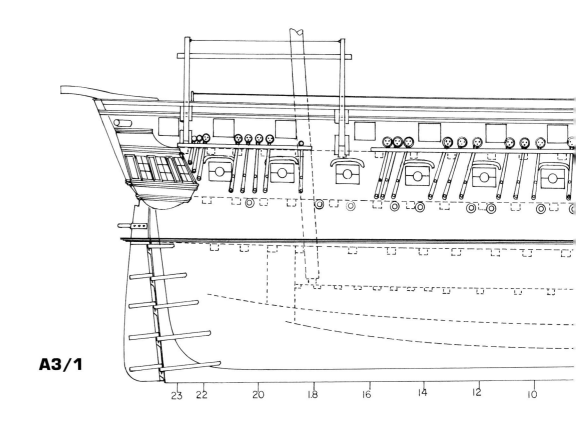

A3/1

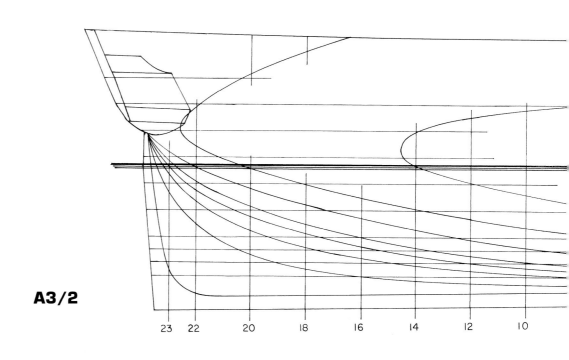

A3/2

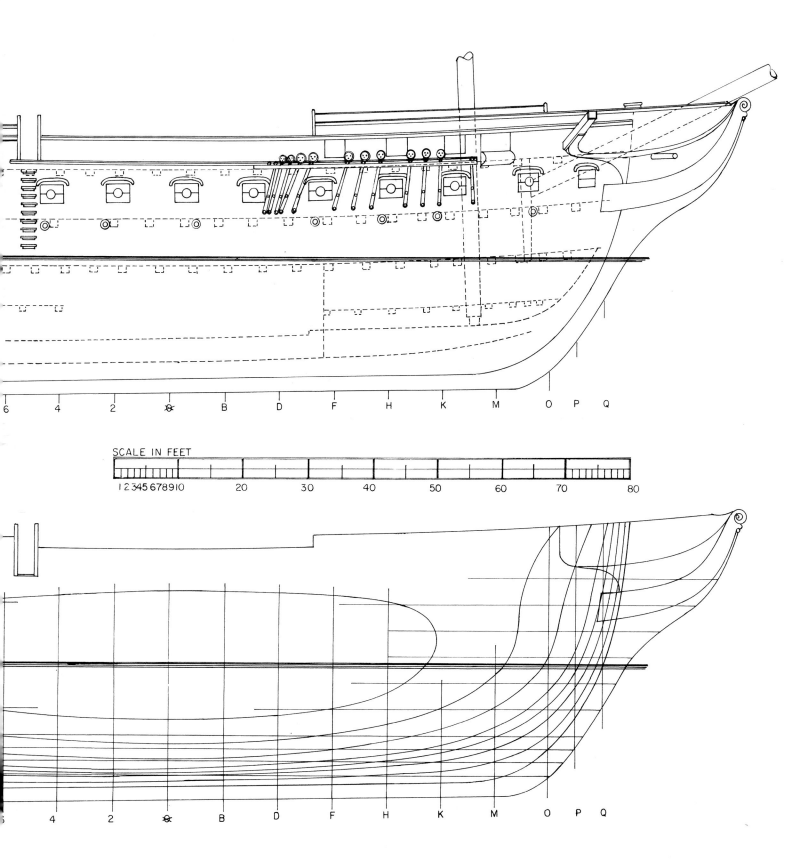

SCALE IN FEET

1 2 3 4 5 6 7 8 9 10   20   30   40   50   60   70   80

6   4   2   ⊕   B   D   F   H   K   M   O P   Q

4   2   ⊕   B   D   F   H   K   M   O P Q

57

A   General Arrangement

A3/3 Half breadth lines

A3/4 Body lines

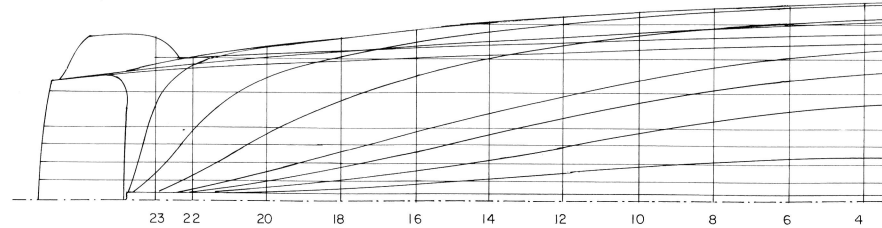

23 22   20   18   16   14   12   10   8   6   4

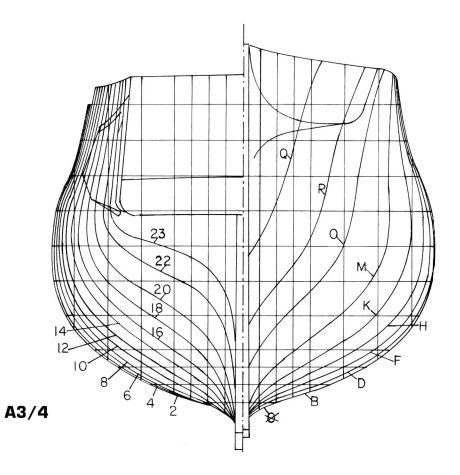

A3/4

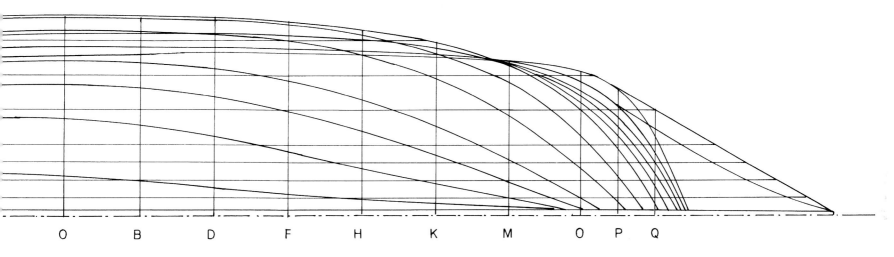

O   B   D   F   H   K   M   O   P   Q

SCALE IN FEET

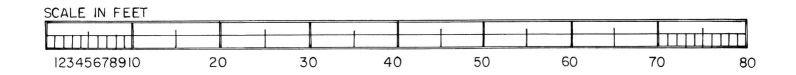

12345678910   20   30   40   50   60   70   80

59

B Hull Structure

B1 KNEES AND RIDERS

1 Diagonal knees
2 Hanging knees
3 Lodging knees
4 Riders

B2 FRAME DISPOSITIONS

1 Stern timbers
2 Transoms
3 Timbers upon the outer stern
  timber
4 Port sill
5 Overhead filling piece
6 Fashion timber
7 Floor timber
8 First futtock
9 Second futtock
10 Third futtock
11 Long top timber
12 Short top timber
13 Filling pieces
14 Hawse pieces
15 Knighthead

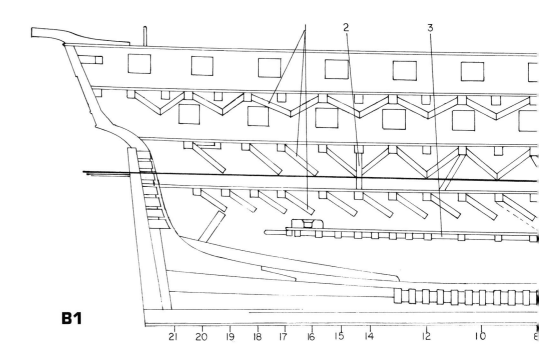

B1

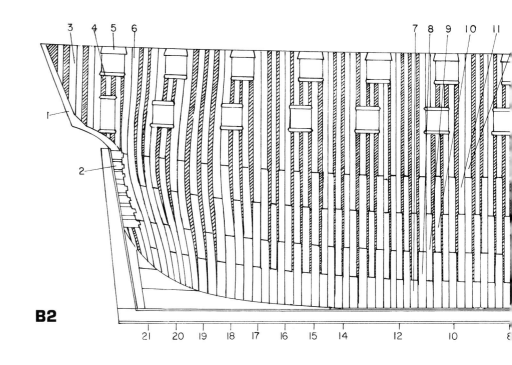

B2

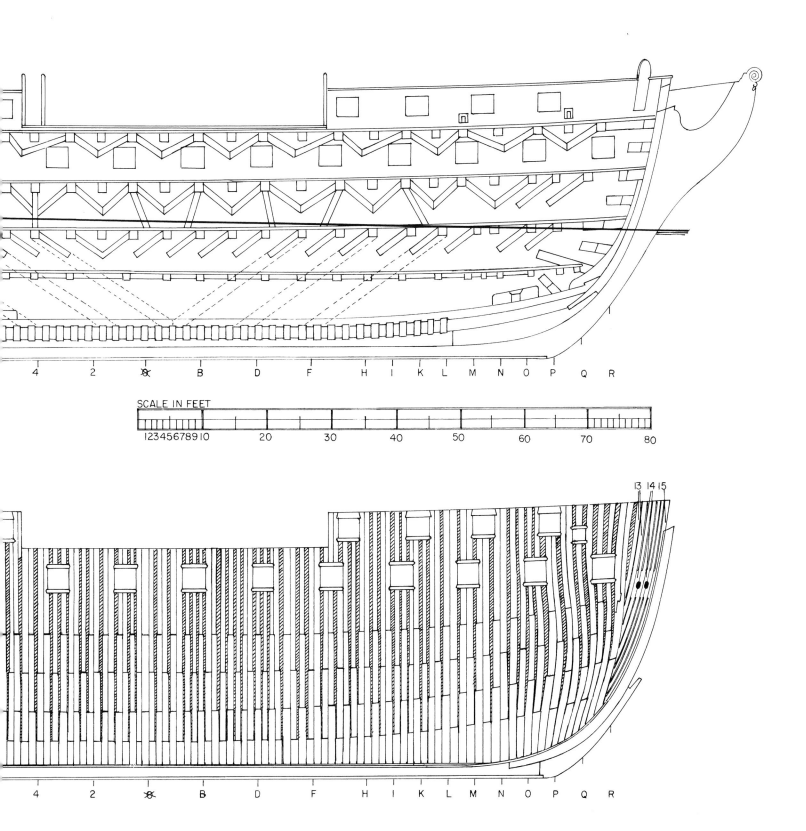

SCALE IN FEET

1 2 3 4 5 6 7 8 9 10   20   30   40   50   60   70   80

B Hull Structure

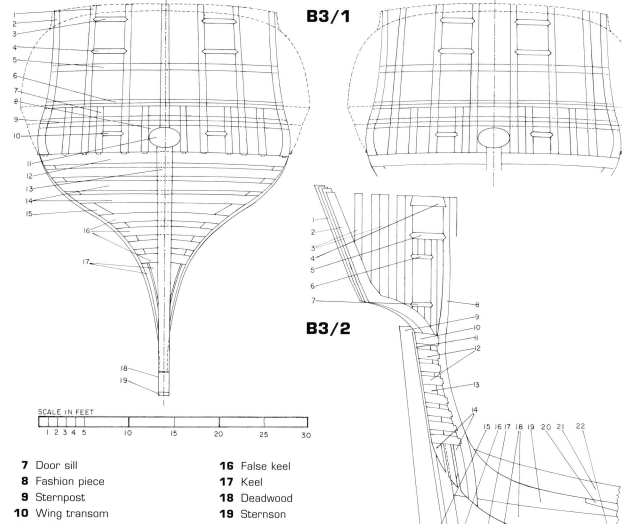

B3/1

B3/2

B3 STERN TIMBERS

B3/1 Six - and five-window proposition rear-view
1 Long stern timber
2 Outer stern timber
3 Upper port sill
4 Port sill
5 Spar-deck transom
6 Window sill transom
7 Short stern timber
8 Helm-port filling pieces
9 Counter transom
10 Loading and airing upper port sill
11 Helm-port
12 Wing transom
13 Sternpost
14 Upper transoms
15 Fashion piece
16 Lower transoms
17 Filling pieces
18 Keel
19 False keel

B3/2 Stern timbers side-view
1 Long stern timber
2 Outer stern timber
3 Timbers upon the outer stern timber
4 Upper port sill filling piece
5 Port sill
6 Upper door sill

7 Door sill
8 Fashion piece
9 Sternpost
10 Wing transom
11 Rabbet
12 Transoms
13 Inner sternpost
14 Filling pieces
15 Iron heel band

16 False keel
17 Keel
18 Deadwood
19 Sternson
20 Keelson scarf
21 Deadwood above sternson/keelson connection
22 Rabbet break

SCALE IN FEET
1 2 3 4 5   10   15   20   25   30

B4 BOW ASSEMBLY

B4/1 From afore
1 Foremost cant frame
2 Filling pieces
3 Bowsprit hole
4 Hawse timbers
5 Hawse holes

6 Stem
7 Knight head

B4/2 From starboard side
1 Foremost cant frame
2 Filling pieces
3 Airing slots between timbers
4 Hawse timbers

5 Hawse holes
6 Stem
7 Knight head
8 Rabbet
9 False keel
10 Keel
11 Deadwood

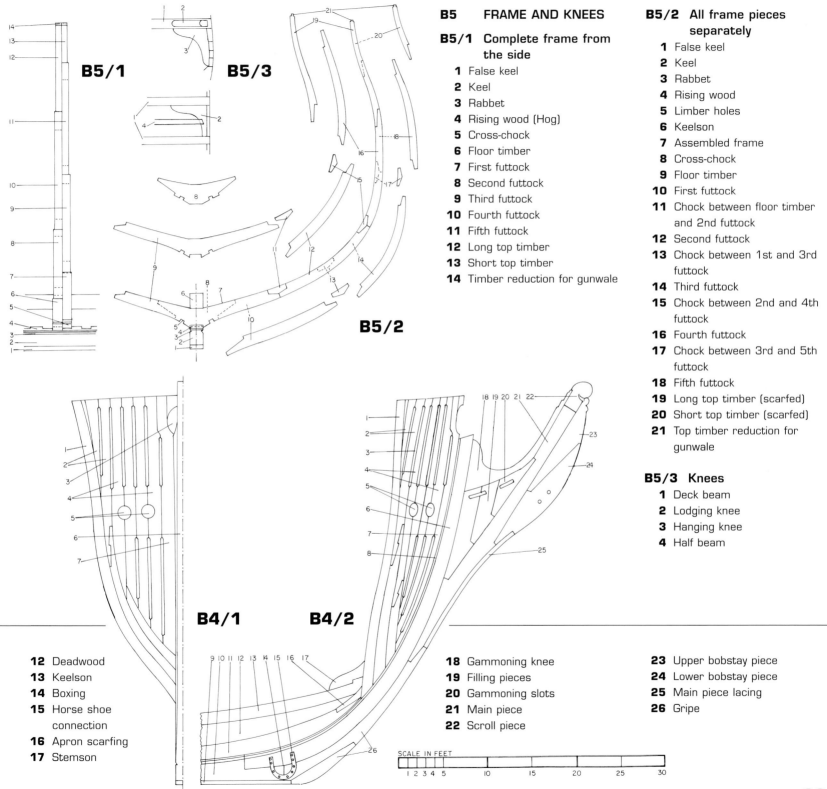

B5/1

B5/3

B5/2

B5   FRAME AND KNEES

B5/1   Complete frame from the side

1   False keel
2   Keel
3   Rabbet
4   Rising wood (Hog)
5   Cross-chock
6   Floor timber
7   First futtock
8   Second futtock
9   Third futtock
10   Fourth futtock
11   Fifth futtock
12   Long top timber
13   Short top timber
14   Timber reduction for gunwale

B5/2   All frame pieces separately

1   False keel
2   Keel
3   Rabbet
4   Rising wood
5   Limber holes
6   Keelson
7   Assembled frame
8   Cross-chock
9   Floor timber
10   First futtock
11   Chock between floor timber and 2nd futtock
12   Second futtock
13   Chock between 1st and 3rd futtock
14   Third futtock
15   Chock between 2nd and 4th futtock
16   Fourth futtock
17   Chock between 3rd and 5th futtock
18   Fifth futtock
19   Long top timber (scarfed)
20   Short top timber (scarfed)
21   Top timber reduction for gunwale

B5/3   Knees

1   Deck beam
2   Lodging knee
3   Hanging knee
4   Half beam

B4/1   **B4/2**

12   Deadwood
13   Keelson
14   Boxing
15   Horse shoe connection
16   Apron scarfing
17   Stemson

18   Gammoning knee
19   Filling pieces
20   Gammoning slots
21   Main piece
22   Scroll piece

23   Upper bobstay piece
24   Lower bobstay piece
25   Main piece lacing
26   Gripe

SCALE IN FEET
1 2 3 4 5   10   15   20   25   30

C External Hull

C1 SIDES

C1/1 Side elevation

1 Captain's gig
2 Stern or Horn davit
3 Rough tree rail
4 Stern timbers
5 Taffrail
6 Stern
7 Main brace spreader
8 Quarter gallery

9 Main course sheet fairlead
10 Carronade port
11 Hammock rail end piece
12 Chained thimble for mizzen topgallant backstay
13 Hammock rail stanchion
14 Hammock rail netting
15 Whale boat
16 Mizzen shroud dead eye
17 Spanker mast
18 Mizzenmast
19 Mizzen channel
20 Dead eye chain
21 Quarter davit

22 Chained thimble for main topgallant backstay
23 Dead eye for main backstay
24 Main channel
25 Drift rail
26 Gunport eyebrow
27 Gunport
28 Hinged half-port lids
29 Mainmast
30 Gunwale
31 Gangway
32 Gangway entrance pieces
33 Fore course sheet fairlead
34 Waist stanchion

35 Waist cloth
36 Commodore's Barge
37 Waist rail
38 Thick stuff above the wales
39 Wales
40 Chess-tree
41 Second bower
42 Chained thimble for fore topgallant backstay
43 Waist end piece
44 Shank painter
45 Fire hearth funnel
46 Fore channel
47 Foremast shroud dead eye

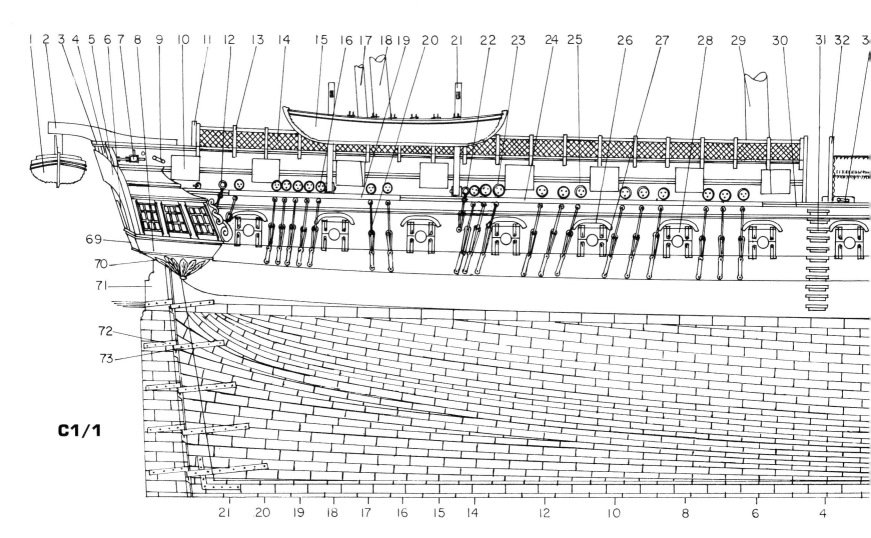

C1/1

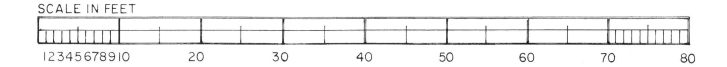

48 Exposed top timber for shank painter to fasten	**53** Cathead	**60** Bowsprit	**67** Knee of the head
49 Sheet anchor	**54** False rail	**61** Scroll head	**68** Copper sheathing
50 Foremast	**55** Entrance to the head	**62** Upper cheek	**69** Upper counter
51 Lashing of anchor stock to exposed top timber	**56** Main rail	**63** Lower cheek	**70** Lower counter
	57 Gammoning	**64** Bobstay holes	**71** Rudder
52 Cathead stopper	**58** Boomkin	**65** Head timbers	**72** Pintle
	59 Lower rail	**66** Naval hood	**73** Gudgeon

SCALE IN FEET

1 2 3 4 5 6 7 8 9 10   20   30   40   50   60   70   80

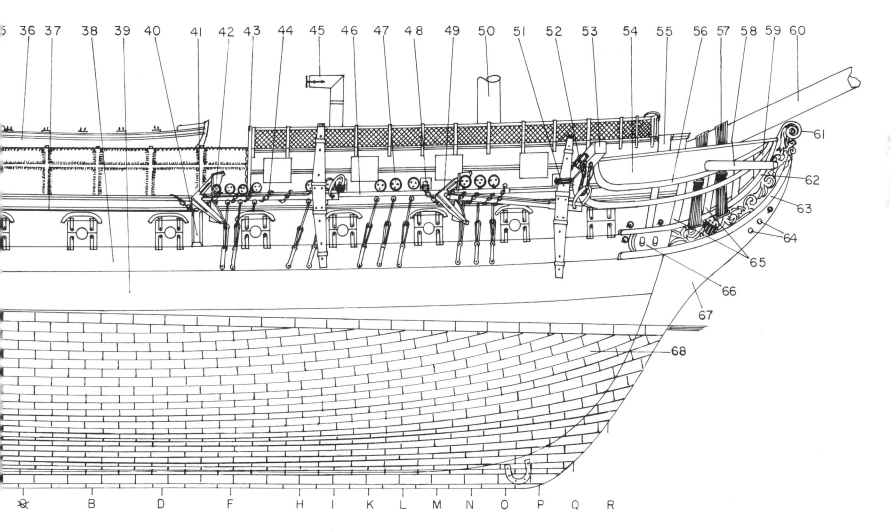

C External Hull

C1/2

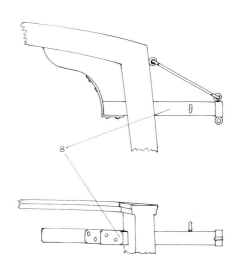

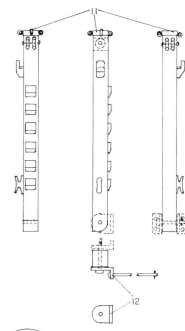

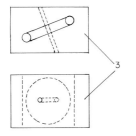

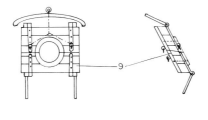

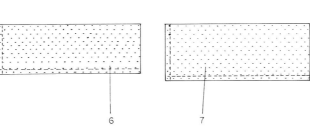

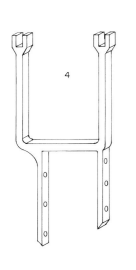

C1/2 Side details

1 Stern davit
2 Spanker boom rest on taffrail
3 Fairlead
4 Hammock rail stanchion
5 Chess-tree
6 Copper side sheath
7 Copper keel sheath
8 Main brace beam (fitted)

9 Gun port lids
10 Opened port lids
11 Quarter davits
12 Quarter davit turning lugs
   (as seen from above)
13 Hammock rail end board
14 Gangway board
15 Gangway step

C2 STERN AND GALLERY

C2/1 Six-window
configuration after
Cornè's painting of
1812

1 Rough tree rail
(closed in)

2 Carronade port

3 Taffrail

4 Cove

5 Window

6 Upper counter

7 Lower finish

8 Lower counter

9 Helm port

10 Lower counter airing ports,
also small item loading ports

C2/2 Five-window
configuration after
Cornè's painting of
1812

C2/3 Cornè's 1803
configuration

C2/4 Current stern
since the 1870s
(approx)

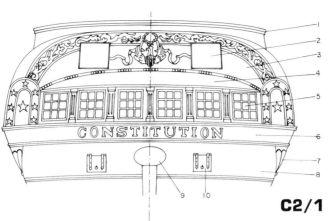

C2/1

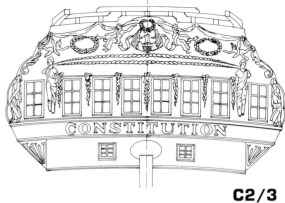

C2/3

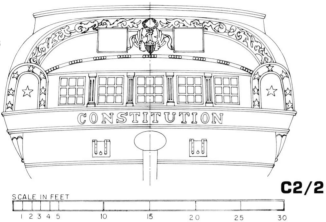

C2/2

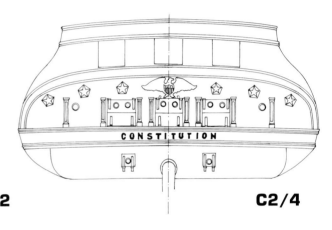

C2/4

SCALE IN FEET
1 2 3 4 5   10   15   20   25   30

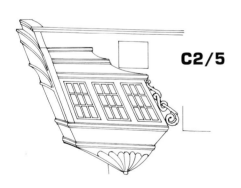

C2/5

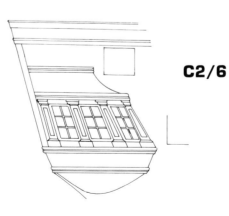

C2/6

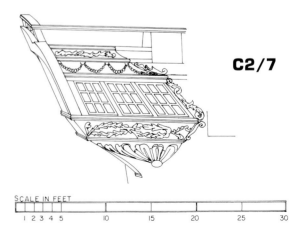

C2/7

SCALE IN FEET
1 2 3 4 5   10   15   20   25   30

C2/5 Gallery as on 1844
draught

C2/6 Current gallery

C2/7 Gallery according to
the earliest 1796
draughts

C External Hull

C3/3

C3 BOW AND HEAD
DETAILS

C3/1 **From afore**

1 Cathead
2 Cathead supporter and
   second head rail
3 Ironbound dead eye
4 Eyebolt
5 Fore channel
6 Dead eye chains
7 Gunport
8 Head timber
9 Upper cheek or hair bracket
10 Naval wood
11 Thickstuff above the wales

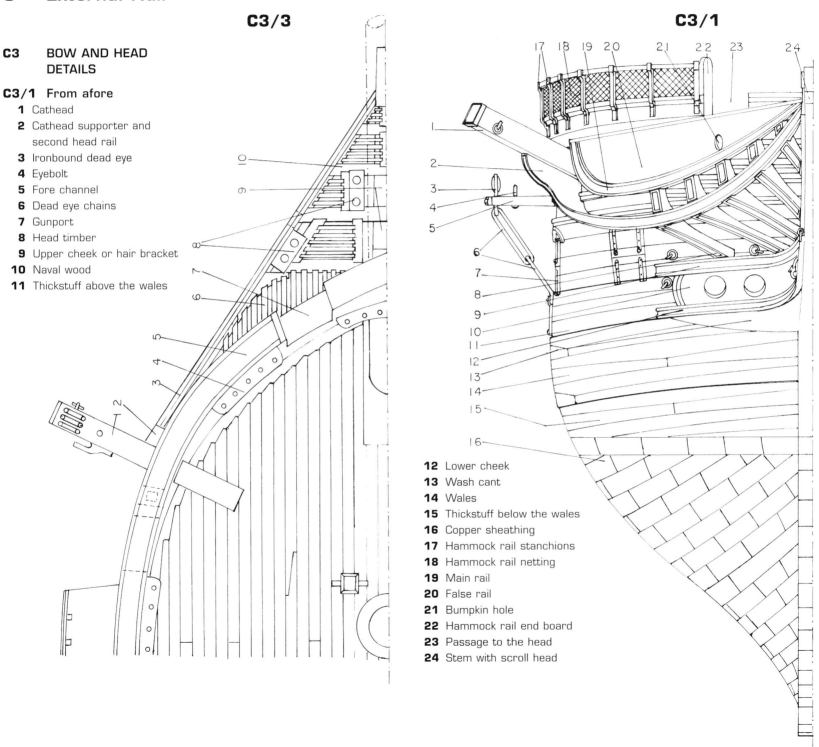

12 Lower cheek
13 Wash cant
14 Wales
15 Thickstuff below the wales
16 Copper sheathing
17 Hammock rail stanchions
18 Hammock rail netting
19 Main rail
20 False rail
21 Bumpkin hole
22 Hammock rail end board
23 Passage to the head
24 Stem with scroll head

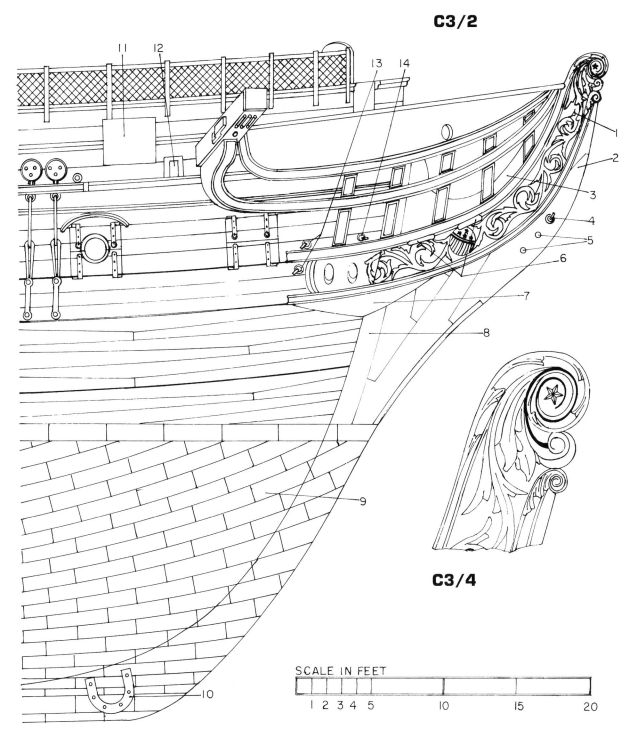

C3/2

C3/2 Side view

1 Trail-board with ornamentation
2 Bobstay piece
3 Gammoning knee
4 Ringbolt for forward bumpkin shroud
5 Bobstay holes
6 Gammoning slots
7 Wash cant
8 Stem
9 Copper sheathing
10 Horseshoe
11 Open gunport
12 Freed top-timber for anchor stowage
13 Ringbolts for bowsprit shrouds
14 Ringbolt for bumpkin after shroud

C3/3 Head

1 Cathead
2 Main rail
3 False rail
4 Pin rail
5 Hammock rail
6 Head gratings
7 Passage to head
8 Seats of ease
9 Gammoning openings
10 Cross timbers

C3/4 Detail of scroll head

C3/4

SCALE IN FEET

1 2 3 4 5   10   15   20

D Internal Hull

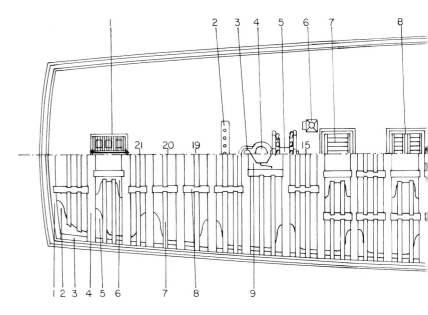

D1 DECKS

D1/1 Spar-deck

1 Cabin skylight
2 Mizzen jeer bits
3 Spanker mast step
4 Mizzenmast opening and surround
5 Steering wheel
6 Binnacle
7 Companion
8 Stair hatchway to wardroom
9 Capstan
10 Aft stair hatchway
11 Main jeer bits
12 Mainmast opening and surround
13 Main topsail sheet bits
14 Rear ledge of spar-deck opening
15 Stairway to gun-deck
16 Portside coaming of spar-deck opening
17 Exposed deck beam
18 Boat chocks
19 Centre gang plank
20 Deck timber doubling above fire hearth
21 Funnel opening
22 Caboose grating
23 Fore jeer bits
24 Foremast opening and surround
25 Fore topsail sheet bits
26 Bowsprit cover-board
27 Bowsprit opening
28 Head

D1/2 Spar-deck framing

1 Spar-deck transom
2 Lodging knee
3 Clamp
4 Deck beam
5 Diagonal hanging knee

6 Double lodging knee
7 Half beam
8 Carling
9 Mizzenmast chocks
10 Capstan chocks
11 Mainmast chocks
12 Lose deck beam for main hatchway clearing
13 Lodging knee
14 Diagonal hanging knee
15 Foremast chocks
16 Spar-deck hook
17 Stem

D1/3 Gun deck

1 Helm port
2 Captain's quarters
3 Sleeping compartment
4 Captain's day cabin
5 Mizzenmast
6 Captain's pantry
7 Stair hatchway to ward room
8 Capstan
9 Aft stair hatchway
10 Double chain pump
11 Crank-handle for chain pumps
12 Mainmast
13 Fore chain pump
14 Main hatchway with stairway to berth deck
15 Fore hatchway with stairway
16 Stairway to berth deck
17 Aft riding bits
18 Fire hearth
19 Fore riding bits
20 Foremast
21 Scuttle
22 Bowsprit step
23 Manger
24 Stem
25 Knee to head

D1/4 Gun deck framing

1 Wing transom
2 Lodging knee
3 Carling
4 Half beam
5 Quadrant
6 Deck beam
7 Centre carling
8 Clamp
9 Mizzenmast chock

10 Capstan chock
11 Iron capstan seating
12 Mainmast chock
13 Beam arm
14 Riding bit post
15 Diagonal hanging knee
16 Foremast chock
17 Breast hook

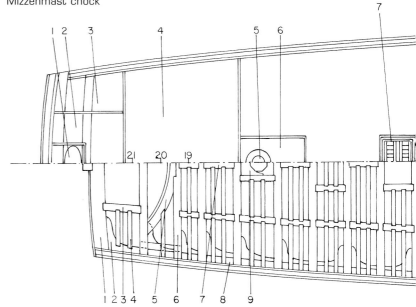

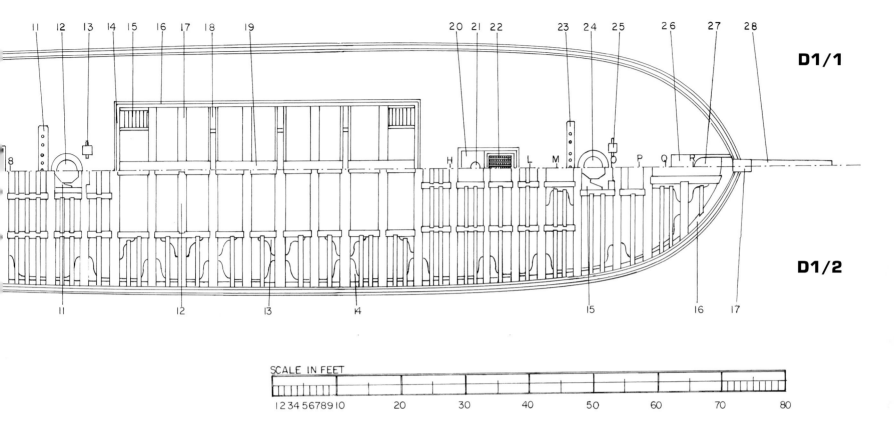

D1/1

D1/2

SCALE IN FEET

1 2 3 4 5 6 7 8 9 10   20   30   40   50   60   70   80

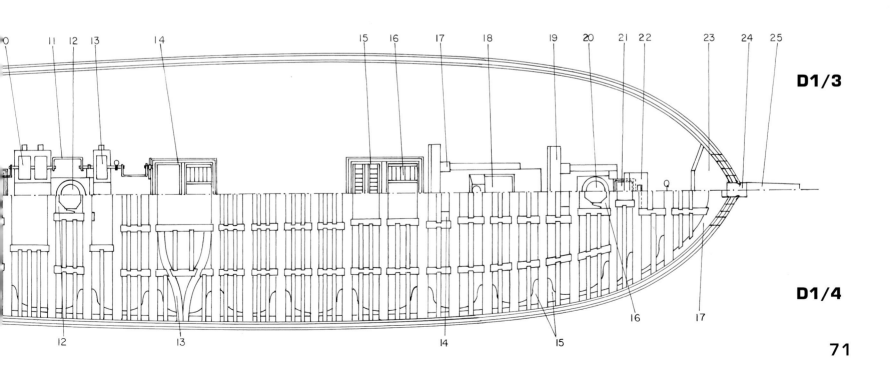

D1/3

D1/4

71

D   Internal Hull

D1/5   Berth deck
1 Sternpost
2 Officers storage
3 Hatchway to bread room
4 Officers pantry
5 Ward room
6 Midshipman's cabin
7 Midshipman's cabin
8 Mizzenmast
9 Junior officers cabin
10 Junior officers cabin
11 Junior officers cabin
12 Junior officers cabin
13 Senior officers cabin
14 Steerage
15 Hatchway
16 Senior warrant officers cabin
17 Stair hatchway
18 Chain pump casings
19 Warrant officer mates cabin
20 Warrant officer mates cabin
21 Mainmast
22 Main hatchway
23 Fore hatchway
24 Brick (Prison)
25 Foremast
26 Sick bay (Hospital)
27 Scuttle
28 Storage
29 Stem

D1/6   Berth deck framing
1 Transom
2 Carling
3 Half beam
4 Deck beam
5 Diagonal hanging knee
6 Clamp
7 Mizzenmast chock
8 Double lodging knee
9 Carlings
10 Chain pump scuttle
11 Mainmast chock

12 Riding bit post
13 Foremast chock
14 Breast hook

D1/7   Orlop deck
1 Sternpost
2 Lowest transom
3 Room stanchion
4 Deadwood above sternson
5 Bread room
6 Mizzenmast step
7 Entrance room with powder
   room scuttle
8 Steward
9 Cockpit
10 Hatchway to purser's wet
   store
11 Surgeon mate's cabin
12 Carpenter walk
13 Slop room
14 Scuttle
15 Mariners clothing store
16 Hatchway to spirit room
17 Purser's room
18 Chain pump casing
19 Mainmast
20 Platform for spare anchors
21 Main hatchway
22 Cable tier stanchions
23 Cable tier
24 Carpenter's walk
25 Fore hatchway
26 Boatswain's store
27 Carpenter's walk
28 Sail room
29 Passage to light room
30 Scuttle to light room
31 Foremast
32 Scuttle
33 Gunner's store
34 Stem

D1/8   Orlop deck framing
16 Filling pieces below transom
17 Lodging knee
18 Filling boards beneath
   mizzenmast step
19 Aft powder magazine scuttle
20 Deck beam
21 Half beams
22 Keelson
23 Mainmast chock
24 Fore powder magazine
   scuttle
25 Foremast chock
26 Breast hook

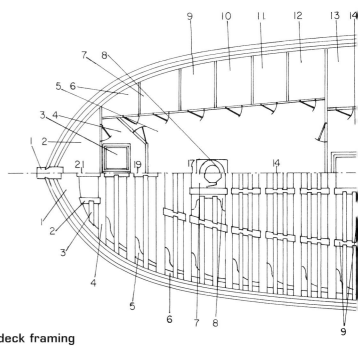

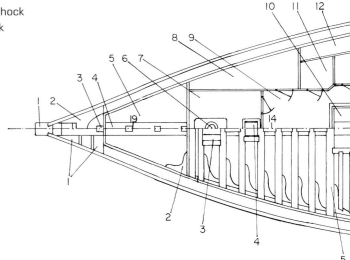

72

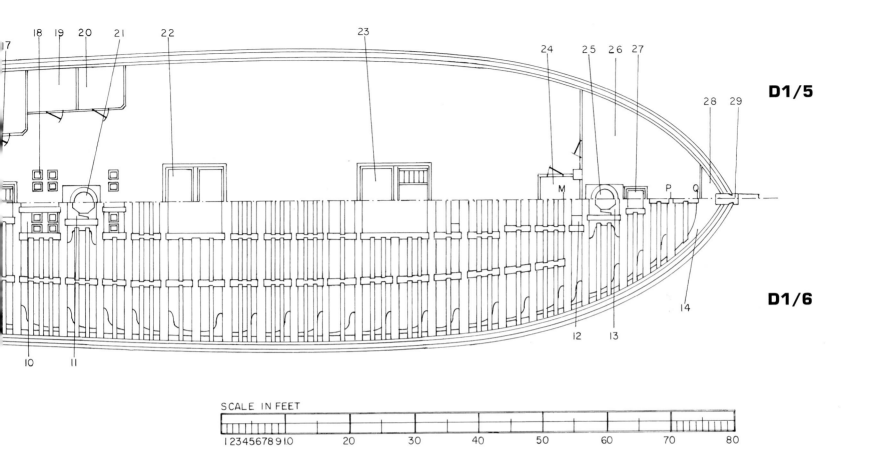

D1/5

D1/6

SCALE IN FEET

1 2 3 4 5 6 7 8 9 10   20   30   40   50   60   70   80

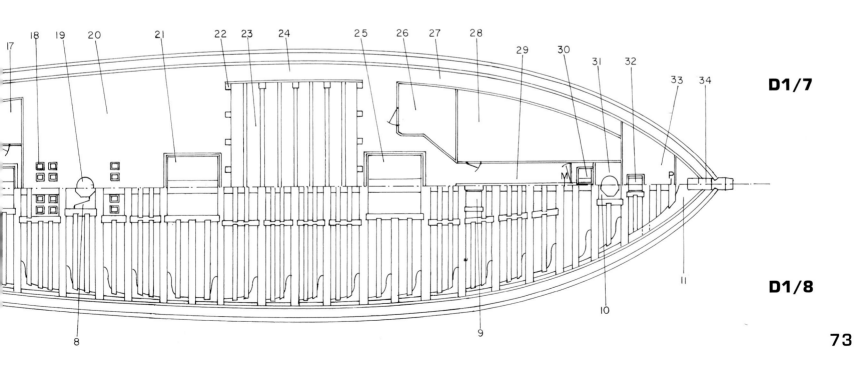

D1/7

D1/8

D Internal Hull

D2 CROSS-SECTIONS

D2/1 Mid-ship cross-section of drawing from 1927

1 Loose keel
2 Keel
3 Garboard strakes
4 Bottom planks
5 Frame
6 Side planks
7 Thick-stuff below the wales
8 Wales
9 Thick-stuff above the wales
10 Lower sill
11 Gunport
12 Upper sill
13 Lower sheer strake
14 Sheer rail
15 Upper sheer strake
16 Gunwale
17 Top timber
18 Waist rail
19 Boats beam
20 Spar deck beam
21 Hanging knee
22 Iron stanchion
23 Gun deck centre carling
24 Half beam
25 Beam
26 Outer carling
27 Lodging knee
28 Hanging knee
29 Timber stanchion
30 Berth deck centre carling
31 Beam
32 Room stanchion
33 Orlop decking 2in
34 Orlop deck beam
35 Deadwood
36 Keelson
37 Counter keel
38 Limber gate
39 Boat hatch carling

40 Hatch ledge
41 Spar deck binding strakes
42 Lodging knee
43 Spar deck planking
44 Chine
45 Waterway
46 Clamps
47 Thick-stuff below clamps
48 Gun deck king planks
49 Inner binding strakes
50 Deck planks
51 Outer binding strakes
52 Deck planks
53 Gun deck chine
54 Waterway
55 Spirketting

56 Clamps
57 Thick-stuff below clamps
58 Berth deck ceiling
59 Standard knee
60 King planks
61 Inner binding strakes
62 Planks
63 Half-beam
64 Outer carling
65 Outer binding strake
66 Planks
67 Diagonal knee
68 Berth deck chine
69 Waterway

70 Clamps
71 Thick-stuff below clamps
72 Diagonal bracing
73 Ceiling
74 Stringer
75 Lodging knee
76 Orlop deck clamps
77 Thick-stuff below clamps
78 Thick-stuff about the floor head
79 Floor ceiling
80 Limber strakes
81 Loose limber plank

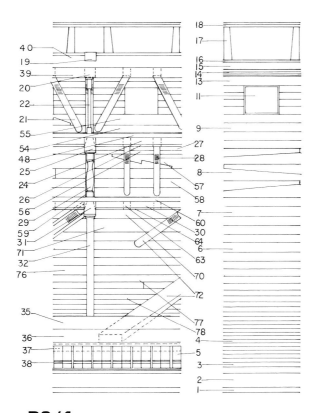

D2/1

SCALE IN FEET

0.5 1 2 3 4 5 6 7 8 9 10   15   20   25   30

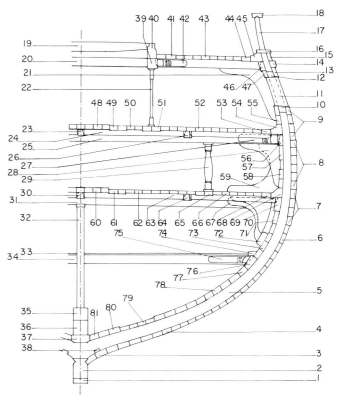

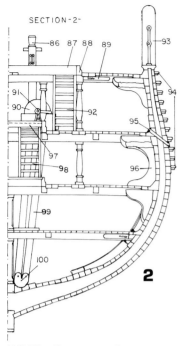
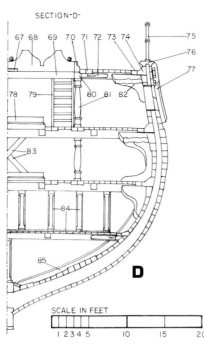
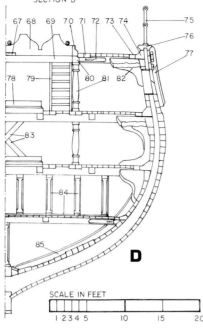

SCALE IN FEET

1 2 3 4 5   10   15   20

D2/2 Cross-sections
2, D, H, O

1 Hammock rail stanchion
2 Plank sheer
3 Sheer planks
4 Drift rail
5 Drift planks
6 Gunwale
7 Waist rail
8 Gun port eye brow
9 Gun port half lids
10 Thick stuff above the wales
11 Wales
12 Thick stuff below the wales
13 Side planks
14 Garboard strakes
15 Keel forefoot
16 False keel
17 Boxing
18 Deadwood
19 Deadwood
20 Keelson
21 Frame
22 Foremast step
23 Foremast surround
24 Scuttle
25 Bowsprit step

26 Dead eye chain
27 Fore channel
28 Iron-bound dead eye
29 Fore topsail sheet bit
30 Funnel
31 Funnel lashing
32 Timber foundation for funnel
33 Binding strakes
34 Carling
35 Normal deck planking
36 Spirketting
37 Pin rail
38 King planks
39 Spar deck beam
40 Diagonal knees
41 Fire hearth
42 Iron stanchion
43 Riding bit post
44 Waterway
45 Clamp
46 Thick stuff below clamp
47 Ceiling
48 Spirketting
49 King planks
50 Binding strakes
51 Stanchion

52 Gun deck beam
53 Ceiling
54 Standard knee
55 Room stanchion
56 Wall to powder magazine scuttle
57 Berth deck beam
58 Diagonal knee
59 Ceiling
60 Diagonal rider
61 Orlop deck beam
62 Keelson
63 Lose limber plank
64 Limber strakes
65 Thick stuff about the floor-head
66 Lodging knee
67 Centre gangplank
68 Boat chock
69 Deck beam across spar deck opening
70 Coaming of spar deck opening
71 Binding strakes
72 Lodging knee
73 Chine
74 Waterway
75 Waist stanchion

76 Gunwale
77 Chess-tree
78 Main hatchway ledge
79 Stairway
80 Carling for spar deck opening
81 Beam stanchion
82 Diagonal knee
83 Stairways
84 Hawse tier with surrounding stanchions
85 Diagonal rider
86 Main topsail sheet bit
87 Spar deck opening end ledge
88 Spar deck opening coaming
89 Lodging knee
90 Chain pump
91 Crank-handle
92 Stairway
93 Gangway end board
94 Gangway
95 Scupper
96 Diagonal standard knee
97 Main hatchway
98 Stairway to berth deck
99 Chain pump casing
100 Basket

D Internal Hull

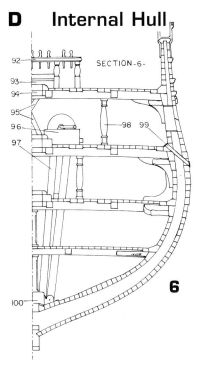

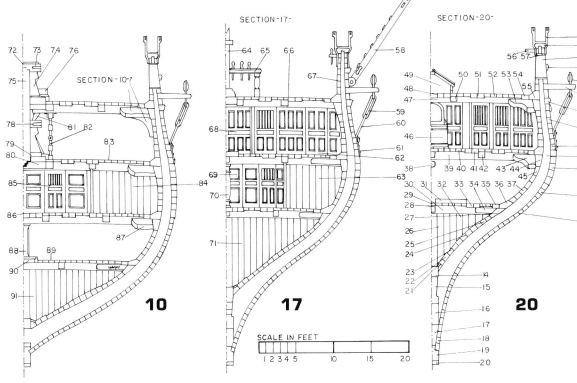

SECTION -6-

6

72 73 74 76
75
78
79
80
81 82
83
85
84
86
87
88 89
90
91

SECTION -10-

10

SECTION-17-

64 65 66
67
68
69
70
71
58
59
60
61
62
63

17

SECTION-20-

49
48
47
46
56 57
50 51 52 53 54
55
39 40 41 42 43 44
45
30 31 32 33 34 35 36 37
38
29
28
27
26
25
24
23
22
21
2
3
4
5
6
7
8
9
10
11
12
13
14
15
16
17
18
19
20

20

SCALE IN FEET
1 2 3 4 5   10   15   20

D2/3 Cross-sections 6, 10, 17, 20

1 Hammock rail stanchion
2 Plank sheer
3 Carronade port
4 Drift rail
5 Iron-bound dead eye
6 Mizzen channel
7 Waist rail
8 Side planks
9 Thick stuff above the wales
10 Frame
11 Wales
12 Thick stuff below the wales
13 Side planks
14 Deadwood above the sternson
15 Sternson
16 Deadwood
17 Deadwood
18 Garboard strakes
19 Keel
20 False keel
21 Lose limber plank
22 Limber strakes
23 Foot-waling

24 Thick stuff about the floor-head
25 Hold ceiling
26 Bread room
27 Clamp
28 Carling
29 Normal deck planking
30 Berth deck beam
31 Hatchway to bread room
32 Bulkhead to lady's hole
33 Binding strakes
34 Lodging knee
35 Chine
36 Waterway
37 Berth deck ceiling
38 Tiller
39 Gun deck beam
40 Normal deck planking
41 Binding strakes
42 Carling
43 Diagonal hanging knee
44 Clamp
45 Thick stuff below clamp
46 Captain's day cabin
47 Spar deck beam
48 King planks

49 Cabin skylight
50 Carling
51 Binding strakes
52 Normal deck planking
53 Diagonal hanging knee
54 Chine
55 Waterway
56 Pin rail
57 Port lining
58 Quarter boat davit
59 Upper chain iron
60 Lower chine iron
61 Chain plate
62 Day cabin bulkhead
63 Officer cabins
64 Spanker mast
65 Mizzen jeer bits
66 Spar deck
67 Spirketting
68 Captain's pantry
69 Ward room
70 Beam stanchion
71 Bread room
72 Capstan
73 Drumhead
74 Whelps

75 Spindle
76 Aft hatchway
77 Diagonal hanging knee
78 Trundle-head
79 Pawl-head
80 Step with iron saucer
81 Stairway
82 Removable stanchion
83 Gun deck
84 Officer's quarter
85 Ward room bulkhead
86 Bert deck
87 Diagonal hanging knee
88 Stanchion
89 Orlop deck
90 Scuttle
91 Spirit room
92 Main jeer bits
93 Aft hatchway stairway
94 Mainmast surrounds
95 Stairway
96 Chain pump's cistern
97 Cain pump's casing
98 Stanchion
99 Scupper
100 Mainmast step

E Fittings

E1 STEERING ARRANGEMENT

E1/1 Side and plan view
1 Rudder stock
2 Tiller
3 Wheel rope tension tackle
4 Gooseneck
5 Quadrant
6 Wheel rope
7 Fairlead
8 Steering wheel
9 Five wheel rope turns around, with the middle nailed to the drum

E1/2 Rudder
1 Rudder head
2 Iron hoops
3 Iron strapping
4 Tiller mortices
5 Upper hance
6 Spectacle plate
7 First after piece (fir)
8 Second hance
9 Second after piece (fir)
10 Rudder stock or helm (oak)
11 Pintle
12 Score in bearding
13 Bearding
14 Filling piece (fir)
15 Bottom piece (elm)
16 Wooden tiller
17 Iron hoop with eyelets to guide wheel rope
18 Gooeseneck
19 Iron hoop with eyelets for tension tackle

E1/3 Steering wheel
1 Especially marked top spoke
2 Felloes
3 Drum or barrel
4 Axel
5 Steering wheel supports

E1/4 Other details
1 Pintles
2 Gudgeons
3 Helmport coat

E1/3

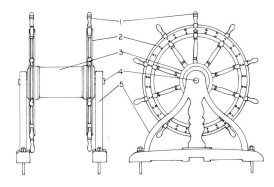

E1/1

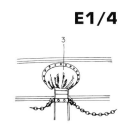

E1/4

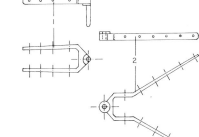

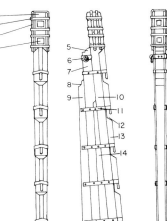

E1/2

E Fittings

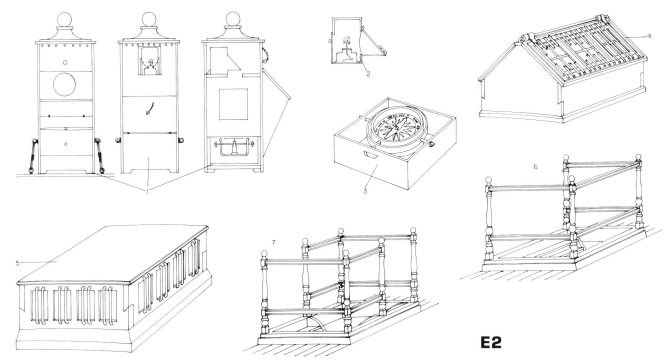

E2 NAVIGATIONAL IMPLEMENTS, SKYLIGHTS AND STAIRWAYS

1 Single binnacle about 1815 by W Burney, LL.D
2 Binnacle oil lamp
3 Gimballed compass in portable wooden box
4 Skylight (Captain's cabin)
5 Skylight (wardroom)
6 Companionway
7 After stairway

E2

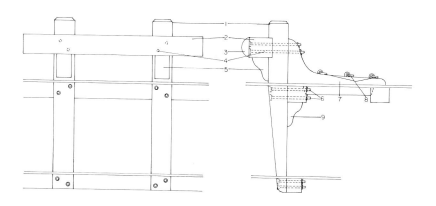

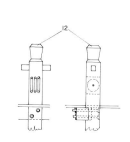

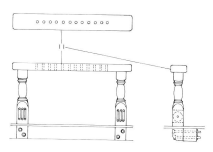

E3

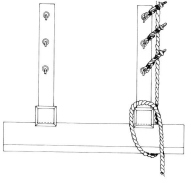

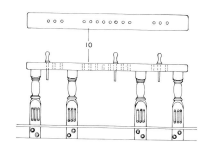

E3 BITTS

1 Bitt pin
2 Crossbeam
3 Soft (fir) crossbeam lining
4 Forelocked bolts
5 Crossbeam supporter
6 Forelocked bolts
7 Standard

8 Ringbolts for cable stoppers
9 Bitt pin supporter
10 Fore and main jeer bitts with belay pins indicated
11 Mizzen jeer bitts
12 Topsail sheet bitts

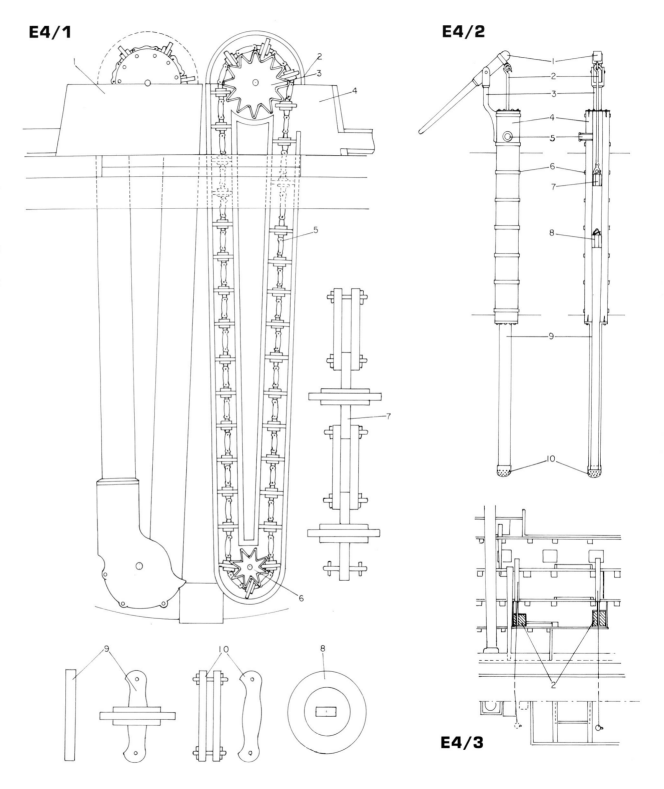

E4/1

E4/2

E4 PUMPS

E4/1 Chain pumps
1 Chain pump casing with opened head
2 Inside a chain pump with head cover indicated
3 Rotating top sprocket wheel
4 Cistern with discharge outlet
5 Water lifting chain
6 Lower sprocket wheel
7 Chain detail
8 Leather saucer as fitted to a chain link (from above)
9 Leather saucer as fitted to a chain link (sideview)
10 Double joining links

E4/2 Elmtree pump
1 Brake
2 Spear
3 Forked brake stanchion
4 Elmtree lifting pipe
5 Discharge
6 Iron hoop
7 Pump shoe
8 Pump bucket
9 Metal suction pipe
10 Basket

E4/3 Deck wash arrangement
(Extract from 1814 Admiralty draft of *President*)
1 Elmtree pump
2 Cistern on orlop deck connected by pipe to outboard

E4/3

79

E Fittings

E5 BRODIE STOVE

1 Baffle plate
2 Turnable funnel

3 Iron hoop with four eyelets to secure the funnel in place
4 Chain drive for roasting spits
5 Roasting spit wheels
6 Drip tray
7 Access to flue
8 Oven door
9 Fireplace door

10 Ashtray door
11 Range grate
12 Front bars
13 Oven
14 Flue
15 Swinging suspension arm
16 Gearbox
17 Smoke jack

18 Impeller
19 Boiler feeding pipe
20 Cooling water overflow (condenser)
21 Cooling water mantle (condenser)
22 Cooling water intake funnel (condenser)
23 Condenced water pipe
24 Boiler
25 Boiler cock
26 Fireplace

27 Ventilation pipe with fan to berth deck
28 Chain wheel for spit drive
29 Hinged flap
30 Spits

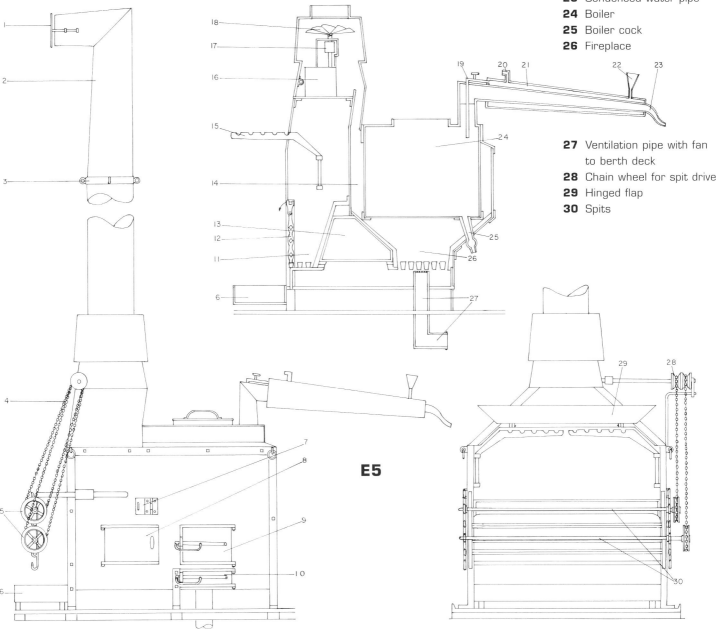

E5

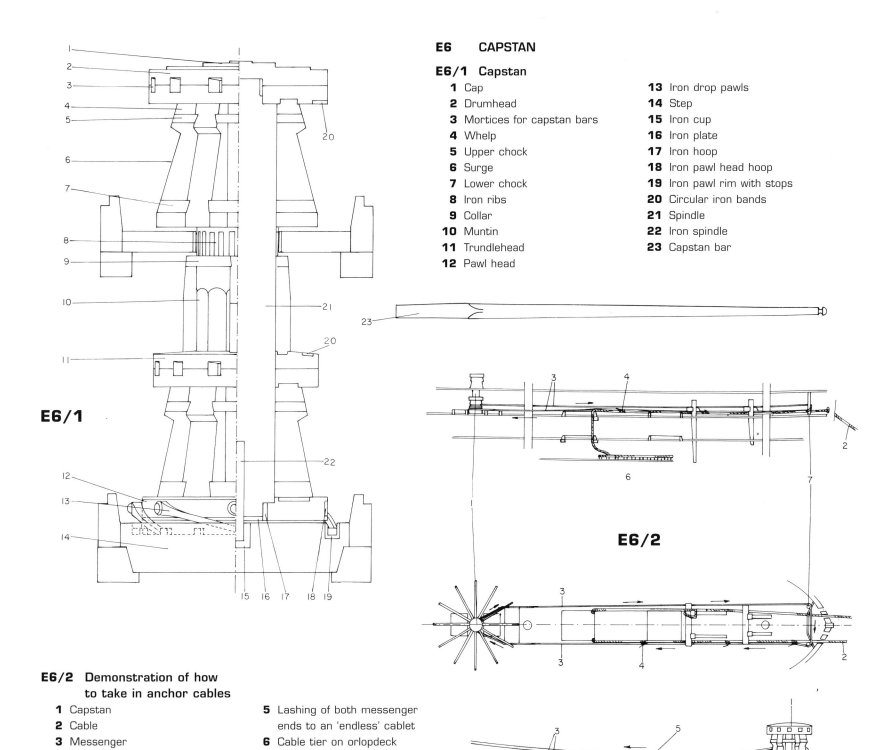

E6 CAPSTAN

E6/1 Capstan

1 Cap
2 Drumhead
3 Mortices for capstan bars
4 Whelp
5 Upper chock
6 Surge
7 Lower chock
8 Iron ribs
9 Collar
10 Muntin
11 Trundlehead
12 Pawl head
13 Iron drop pawls
14 Step
15 Iron cup
16 Iron plate
17 Iron hoop
18 Iron pawl head hoop
19 Iron pawl rim with stops
20 Circular iron bands
21 Spindle
22 Iron spindle
23 Capstan bar

E6/1

E6/2

E6/2 Demonstration of how to take in anchor cables

1 Capstan
2 Cable
3 Messenger
4 Temporary nipper connection of messenger and cable
5 Lashing of both messenger ends to an 'endless' cablet
6 Cable tier on orlopdeck
7 Messenger roller

E   Fittings

E7   ANCHORS AND FASTENINGS

1   Bower or sheet anchor
- A   Ring
- B   Nut
- C   Shank
- D   The point to establish the width of arms
- E   Palm or fluke
- F   Arm
- G   Crown
- H   Blade
- J   Snape
- K   Bill
- L   Throath
- M   Hoop
- N   Stock

2   Stream anchor
3   Kedge anchor
4   Puddening of a ring
5   Cable clich on ring
6   Fisterman's bend on smaller ring
7   Cork buoy: lower end = buoy rope; upper end = short lanyard
8   Anchor with buoy rope connected

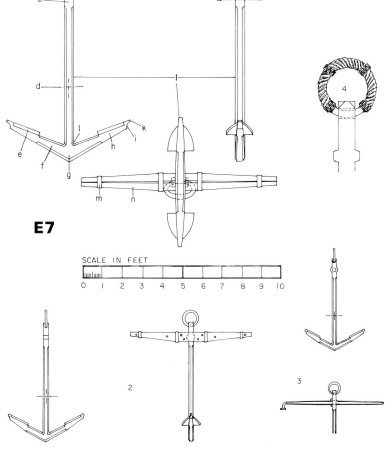

E7

SCALE IN FEET

0   1   2   3   4   5   6   7   8   9   10

E8/2

E8   ANCHOR STOWAGE

E8/1   Cathead

1   Cathead
2   Belay cleat
3   Cathead stopper cleat
4   Cathead stopper hole

5   Cat-tackle sheave
6   Axe with ring
7   Iron hoop

E8/2   Side elevation and plan view

1   Sheet anchor
2   Shank-painter
3   Selvage
4   Fore channel
5   Best bower
6   Cathead stopper
7   Cathead

E8/1

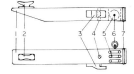

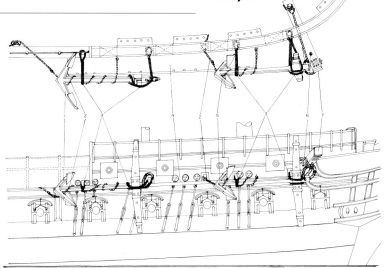

F   Armament

F1   24-PDR LONG GUN

1 24-pdr English long gun
2 18-pdr bow chaser of USS Constitution bored out to take 24-pdr shot
3 Cascable
4 1st Reinforcement
5 2nd Reinforcement
6 Chase
7 Muzzle
8 Muzzle moulding
9 Swell of muzzle
10 Muzzle astragal
11 Bore
12 Breech ring and ogee
13 3rd Reinforce ring and ogee
14 Trunnion
15 2nd Reinforce ring and ogee
16 Chamber
17 1st Reinforce rig and ogee
18 Vent
19 Pan
20 Base ring and ogee
21 Breech
22 Butt
23 Round shot

F1/2   Gun carriage

1 Bracket
2 Stool bed
3 Gun-tackle eyebolt
4 Gun-tackle eyebolt
5 Quadrant
6 Bracket bolts
7 Cap square joint bolt
8 Cap square
9 Swell of muzzle
10 Cap square key
11 Transom
12 Linchpin
13 Axtree
14 Axtree hoop
15 Fore truck
16 Axtree stay
17 Cap square eyebolt
18 Breeching ring-bolt
19 Bed bolt
20 Bolster
21 Rear axtree
22 Rear truck
23 Train-tackle eyebolt

No scale

A Flexible sponge and rammer
B 24-pdr long gun mounted on carriage with breeching, gun tackle and train tackle.

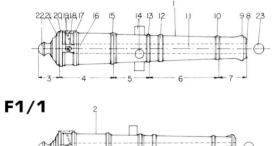

F1/1

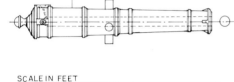

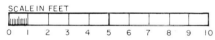

SCALE IN FEET
0 1 2 3 4 5 6 7 8 9 10

F1/2

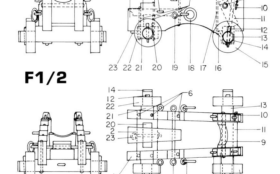

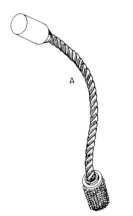

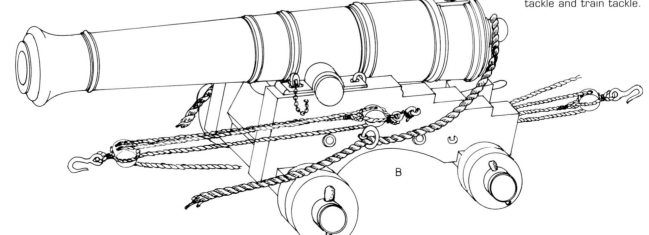

F Armament

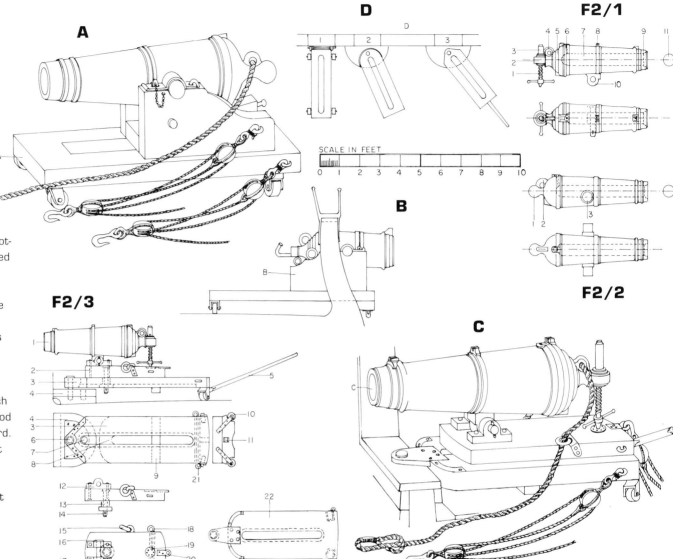

F2 CARRONADES

A A carriage-mounted carronade, with breeching, carriage and slide tackles, as installed on spar deck of Constitution in 1927 and remaining to this day.

B Early carriage-mounted carronade as used until about 1790

C English-style carronade, pivot-mounted and commonly used from about 1795, with breeching, slide or skeat tackle and train tackle. (The carriage tackle is omitted.)

D Different mounting of slides or skeats.

   1 Slide, from approx 1790–95, with a square head and four wheels, which was secured with an iron rod to eyebolts in slide and board.

   2 Rounded head with pivot bolt and two wheels, dated after 1795.

   3 Most common slide post 1795, with angled head, pivot bolt and two wheels.

F2/1 English-style 36-pdr carronade

   1 Butt
   2 Elevating screw box
   3 Leather cover
   4 Breeching ring
   5 Breech
   6 Base patch with pan
   7 Bore
   8 Step sight
   9 Muzzle
   10 Lug
   11 Round shot

F2/2 Carronade from 1927

   1 Elevating screw
   2 Breeching ring
   3 Trunnion

F2/3 Assembled carronade

   1 Carronade
   2 Carriage
   3 Slide or skeat
   4 Pivoting base

   5 Turn spike
   6 Pivot bolt
   7 Second pivoting base bolthole for turning the slide sidewise when not in action
   8 Carriage traversing slot
   9 Strengthening timber
   10 Rear wheel
   11 Spike socket
   12 Joint chock

   13 Gudgeon
   14 Carriage mounting & slide bolt
   15 Breeching ring
   16 Lug bolt
   17 Chock fastening bolts
   18 Carriage tackle eyebolt
   19 Elevating screw plate
   20 Train tackle eyebolt
   21 Slide tackle eyebolt
   22 Alternative pivoting slide/skeet

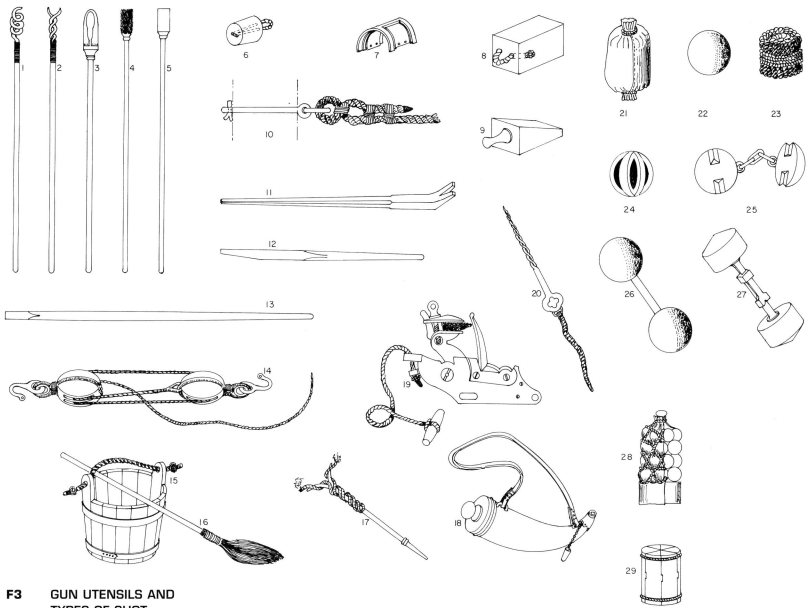

F3   GUN UTENSILS AND TYPES OF SHOT

1 Worm
2 Another type of worm
3 Ladle
4 Sponge
5 Rammer
6 Tompion
7 Apron, weather-cover from lead or copper over pan and vent hole

8 Quoin underlay
9 Quoin
10 Breeching, as seized to a ring-bolt next to a gun port
11 Iron crowbar
12 Gun training spike
13 Slide training spike
14 Gun-tackle, port- or train tackle

15 Half tub for gun cooling
16 Swabber
17 Lint stock
18 Powder horn
19 Gunlock as developed 1780 by Captain Sir Charles Douglas
20 Prime wire
21 Powder cartridge bag

22 Round shot
23 Wad
24 Incendiary round shot (carcass)
25 Chain shot
26 Bar shot
27 Another type of bar shot
28 Case shot
29 Another type of case shot

G Ship's Boats

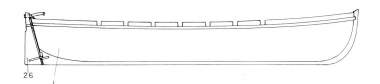

G1

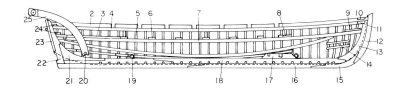

G2/1

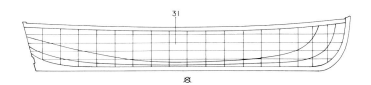

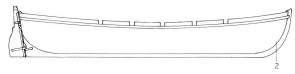

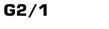

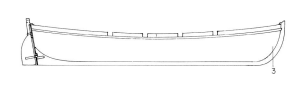

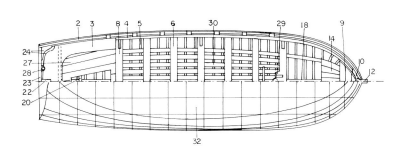

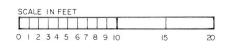

SCALE IN FEET

0 1 2 3 4 5 6 7 8 9 10   15   20

G1   34FT LAUNCH

1 Sheer elevation	**10** Wash-strake breast-hook	**19** Eyebolts	**27** Stern foot-waling
2 Wash-strake with row locks	**11** Breast-hook	**20** Davit footing	**28** Flagpole bracket
3 Stern sheet bench	**12** Stem	**21** Sternson knee	**29** Mast bracket
4 Gunwale	**13** Apron	**22** Sternpost	**30** Open foot-waling
5 Framing	**14** Breast foot-waling	**23** Transom	**31** Vertical bodylines
6 Loose thwarts	**15** Forefoot	**24** Transom knees	**32** Waterlines
7 Rising	**16** Keel	**25** Davit	**33** Cross body lines
8 Fixed thwarts with knees	**17** Mast step	**26** Rudder with tiller (unshipped	**34** Centre cross-section
9 Breast decking	**18** Keelson	when davit in use)	**35** Stern and bow view

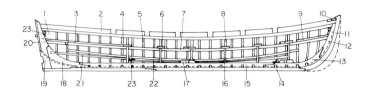

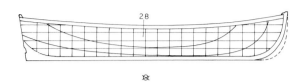

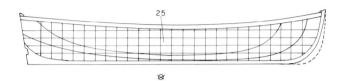

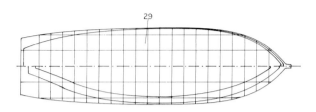

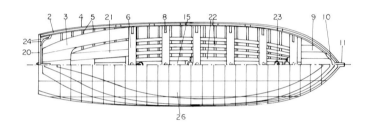

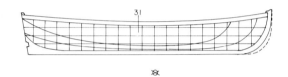

G2/2

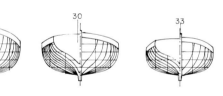

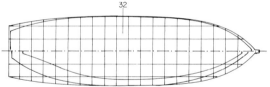

SCALE IN FEET

0 1 2 3 4 5 6 7 8 9 10   15   20

G2   CUTTERS

G2/1   Cutter sheer elevation
1 32ft Cutter (with clincher planking shown)
2 28ft Cutter
3 25ft Cutter

G2/2   Cutter details
1 32ft Cutter profile
2 Wash-strake with row locks
3 Stern sheet bench

4 Gunwale
5 Frames
6 Thwarts
7 Rising
8 Knees
9 Breast decking
10 Breast-hook
11 Stem
12 Apron
13 Forefoot of keelson

14 Keel
15 Keelson
16 Thwart stanchion
17 Mast step
18 Sternson knee
19 Sternpost
20 Transom
21 Stern foot-waling
22 Loose foot-waling
23 Ringbolts

24 Transom knees
25 Vertical body lines
26 Waterlines
27 Cross body lines
28 28ft Cutter vertical lines
29 Waterlines
30 Cross body lines
31 25ft Cutter vertical lines
32 Waterlines
33 Cross body lines

G Ship's Boats

G3

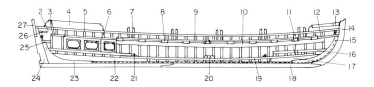

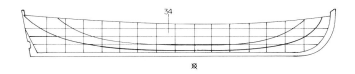

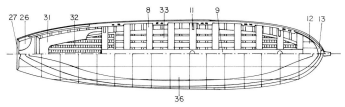

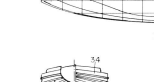

G4

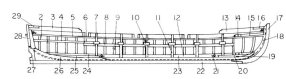

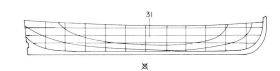

G3   32FT COMMODORE'S BARGE

1 Sheer elevation
2 Profile
3 Back of stern sheet bench
4 Quarter washboard
5 Gunwale
6 Frames
7 Tholes
8 Thwarts
9 Centre plank
10 Rising
11 Thwart knees
12 Breast decking
13 Breast-hook
14 Stem
15 Apron
16 Bow grating with foremast step
17 Deadwood
18 Keel
19 Keelson
20 Mainmast step
21 Ringbolts
22 Stern deadwood
23 Sternpost knee
24 Sternpost
25 Floor
26 Coxwain's seat
27 Transom
28 Rudder
29 Pintle
30 Gudgeon
31 Stern sheet bench
32 Stern sheet grating
33 Loose foot-waling
34 Vertical body lines
35 Cross body lines
36 Waterlines

SCALE IN FEET
0 1 2 3 4 5 6 7 8 9 10   15   20

G4   CAPTAIN'S 25FT GIG

1 Sheer elevation
2 Profile
3 Quarter washboard
4 Stern sheet bench
5 Gunwale
6 Row locks
7 Thwart knees
8 Rising
9 Frames
10 Thwarts
11 Thwart stanchions

12 Loose foot-waling
13 Bow washboard
14 Mast thwart
15 Breast decking
16 Breast-hook
17 Stem
18 Apron
19 Deadwood
20 Mast step
21 Ringbolts
22 Keel
23 Keelson

24 Stern sheet foot-waling
25 Deadwood
26 Sternpost knee
27 Sternpost
28 Transom
29 Transom hook
30 Rudder
31 Vertical body lines
32 Waterlines
33 Cross body lines
34 Clincher planking indicated on stern and front section

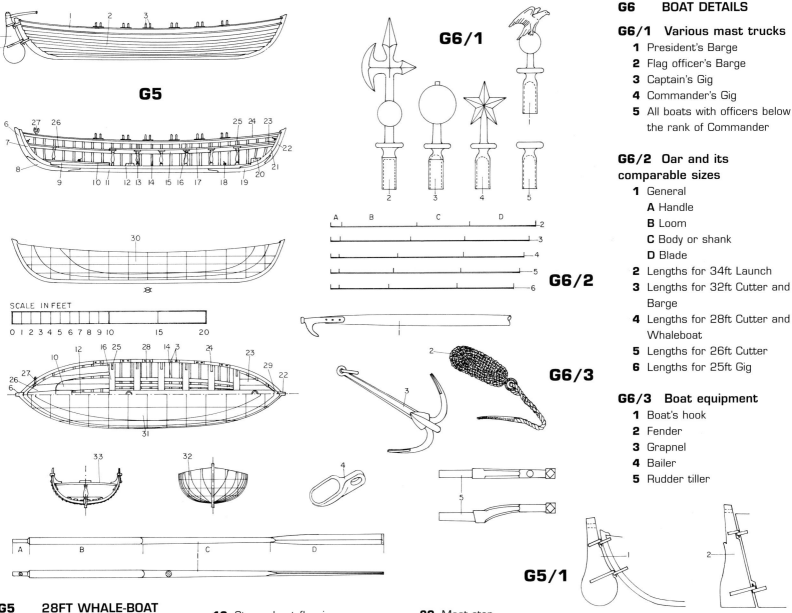

G5

G6 BOAT DETAILS

G6/1 Various mast trucks
1 President's Barge
2 Flag officer's Barge
3 Captain's Gig
4 Commander's Gig
5 All boats with officers below
  the rank of Commander

G6/2 Oar and its comparable sizes
1 General
  A Handle
  B Loom
  C Body or shank
  D Blade
2 Lengths for 34ft Launch
3 Lengths for 32ft Cutter and
  Barge
4 Lengths for 28ft Cutter and
  Whaleboat
5 Lengths for 26ft Cutter
6 Lengths for 25ft Gig

G6/3 Boat equipment
1 Boat's hook
2 Fender
3 Grapnel
4 Bailer
5 Rudder tiller

SCALE IN FEET
0 1 2 3 4 5 6 7 8 9 10   15   20

G5 28FT WHALE-BOAT
1 Sheer elevation
2 Clincher planking
3 Tholes
4 Rudder
5 Pintles and Gudgeons
6 Sternpost
7 Inner sternpost
8 Stern foot
9 Deadwood
10 Stern sheet flooring
11 Keel
12 Keelson
13 Thwart stanchion
14 Frames
15 Rising
16 Thwart
17 Centre plank
18 Ringbolts
19 Forefoot
20 Mast step
21 Apron
22 Stem
23 Breast decking
24 Gunwale
25 Sheer plank
26 Stern sheet bench
27 Swivel thole
28 Foot-waling
29 Breast-hook

30 Vertical body lines
31 Waterlines
32 Cross body lines
33 Cross-section

G5/1 Rudder variants
1 For bent sternpost
2 For straight sternpost

89

G Ship's Boats

G7 STOWAGE OF BOATS

G7/3

G7/1 On deck

1 34ft launch
2 32ft Commodore's barge
3 32ft cutter
4 28ft cutter
5 Fender to safeguard the nesting boat
6 One-sided gripe
7 Boat cradle
8 Two-ended grip

G7/2 In quarter davits

1 Whaleboat
2 Quarter davit
3 Davit lugs
4 Puddened gripe
5 Davit guy
6 Davit head
7 Boat tackle
8 Jackstay, topping lift and mast pendant to lift or lower the quarter davits

G7/3 In stern davits

1 26ft cutter
2 Boat tackle
3 Stern davit
4 Puddened gripe
5 Gripe stern-fast

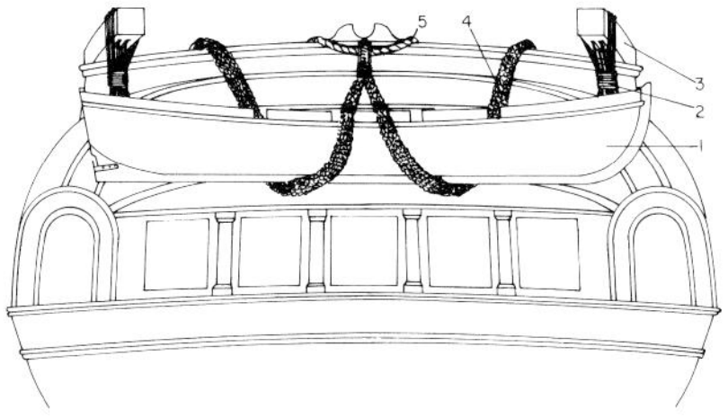

G7/1

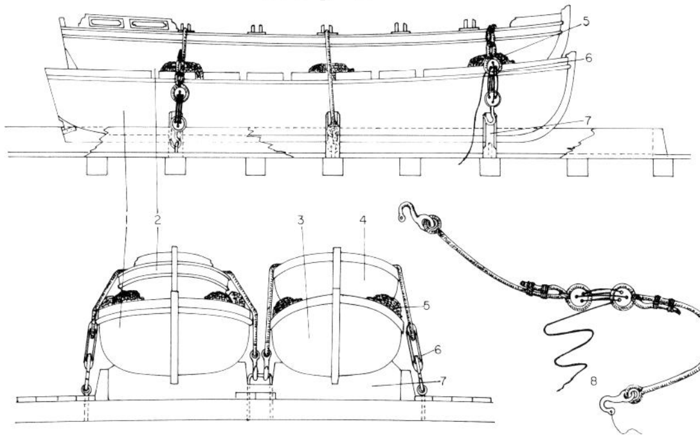

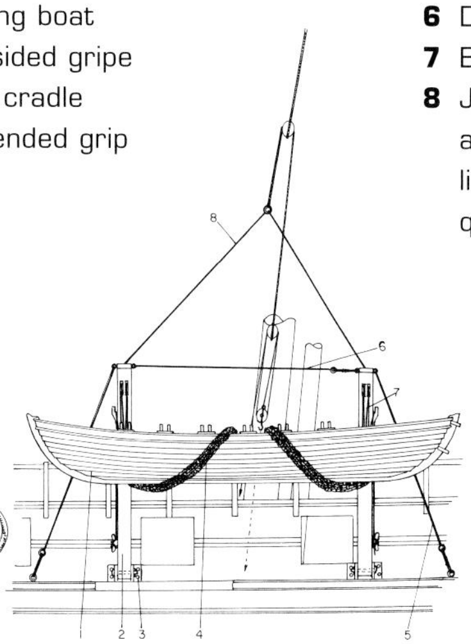

G7/2

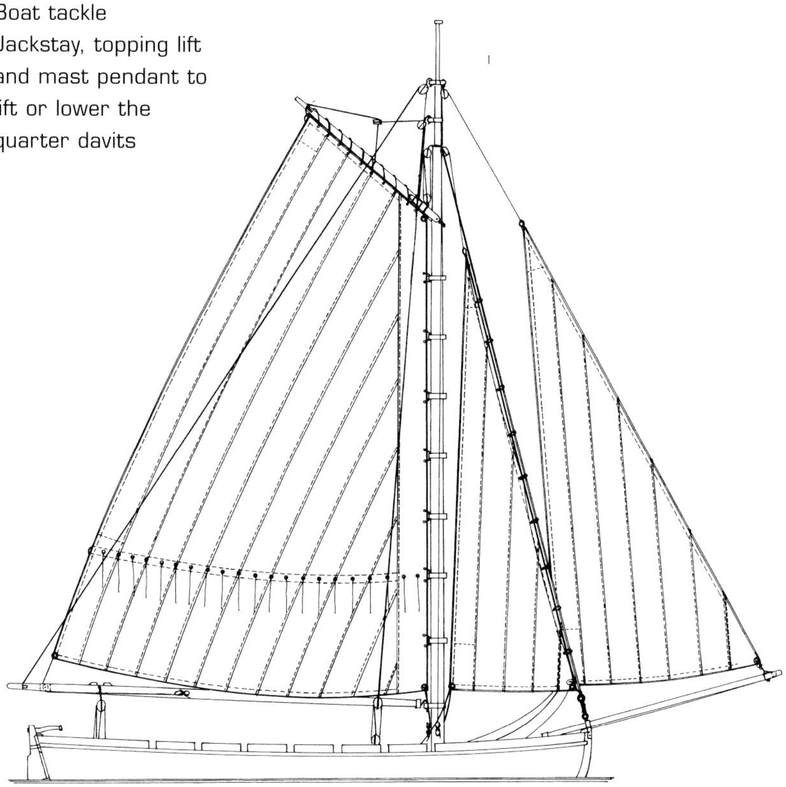

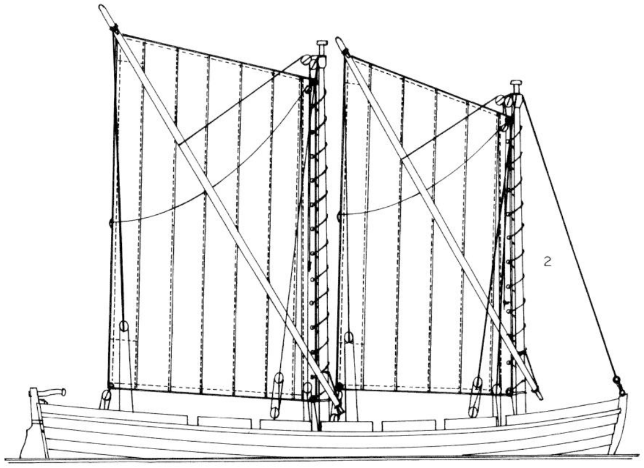

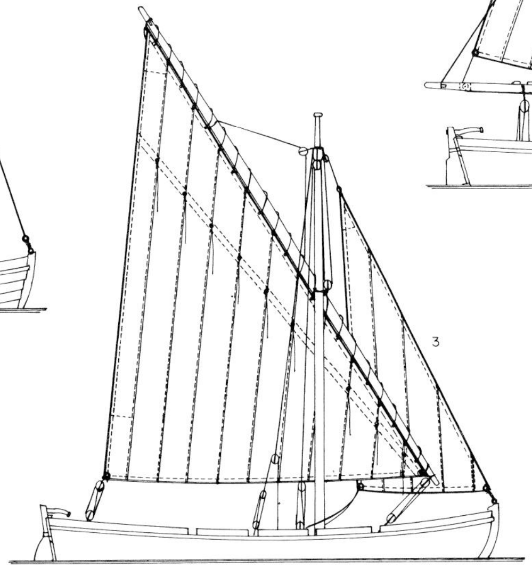

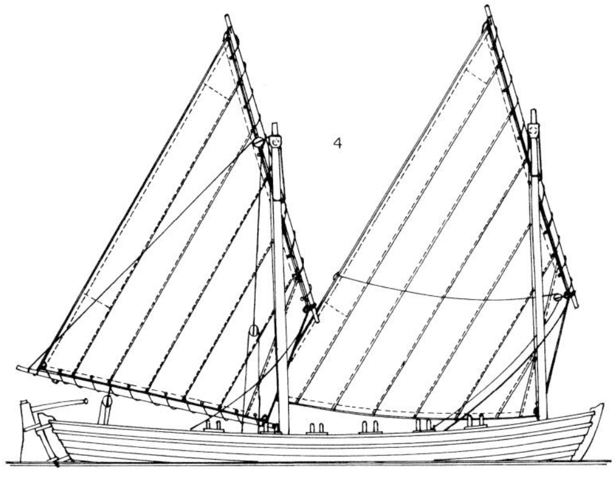

G8 BOAT RIGGINGS

1 Sloop-rigged launch
2 Two-masted sprit-rigged cutter
3 Lateen-rigged cutter
4 Two-masted lug-rigged whaleboat

G8

H Masts and Spars

H1 FULLY RIGGED SHIP
AS BETWEEN 1803
AND 1815
(1815 LISTING)

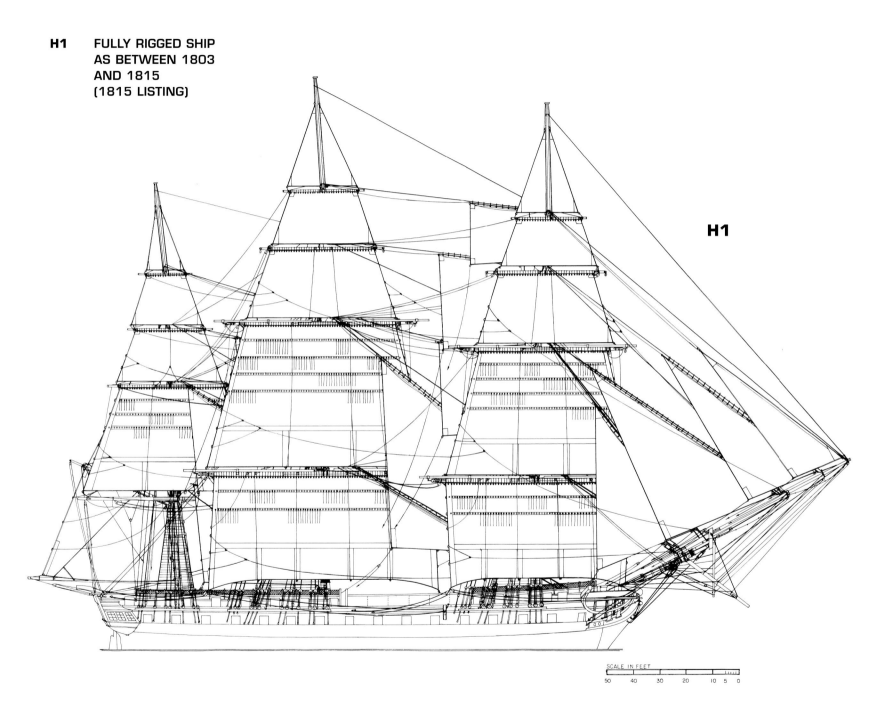

H1

SCALE IN FEET
50   40   30   20   10 5 0

H Masts and Spars

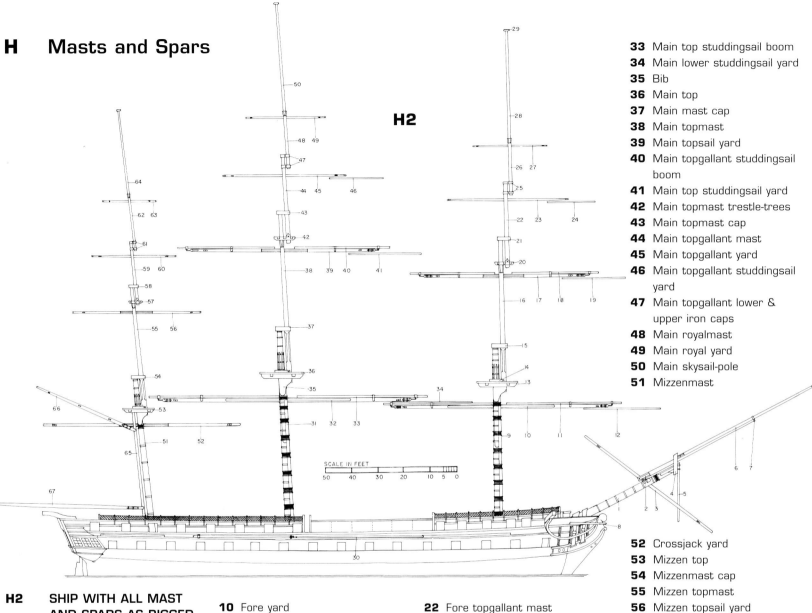

H2

SCALE IN FEET
50 40 30 20 10 5 0

H2 SHIP WITH ALL MAST AND SPARS AS RIGGED BETWEEN 1803 AND 1815 (1815 LISTING)

1 Bowsprit
2 jib boom
3 Spritsail yard
4 Bowsprit cap
5 Martingale or Dolphin striker
6 Flying jib boom
7 Iron jib boom cap
8 Bumpkin
9 Foremast
10 Fore yard
11 Fore topsail studdingsail boom
12 Fore lower studdingsail yard
13 Fore top
14 Bolster
15 Foremast cap
16 Fore topmast
17 Fore topsail yard
18 Fore topgallant studdingsail boom
19 Fore topsail studdingsail yard
20 Fore topmast trestle-tree
21 Fore topmast cap
22 Fore topgallant mast
23 Fore topgallant yard
24 Fore topgallant studdingsail yard
25 Fore topgallant lower & upper iron caps
26 Fore royalmast
27 Fore royal yard
28 Fore skysail-pole
29 Truck
30 Swinging studdingsail boom
31 Mainmast
32 Main yard

33 Main top studdingsail boom
34 Main lower studdingsail yard
35 Bib
36 Main top
37 Main mast cap
38 Main topmast
39 Main topsail yard
40 Main topgallant studdingsail boom
41 Main top studdingsail yard
42 Main topmast trestle-trees
43 Main topmast cap
44 Main topgallant mast
45 Main topgallant yard
46 Main topgallant studdingsail yard
47 Main topgallant lower & upper iron caps
48 Main royalmast
49 Main royal yard
50 Main skysail-pole
51 Mizzenmast

52 Crossjack yard
53 Mizzen top
54 Mizzenmast cap
55 Mizzen topmast
56 Mizzen topsail yard
57 Mizzen topmast trestle-trees
58 Mizzen topmast cap
59 Mizzen topgallant mast
60 Mizzen topgallant yard
61 Mizzen topgallant lower & upper iron caps
62 Mizzen royalmast
63 Mizzen royal yard
64 Mizzen skysail-pole
65 Spanker mast
66 Spanker gaff
67 Spanker boom

H2/1 Masts

1 Mainmast, front
2 Mainmast, side
3 Foremast, rear
4 Foremast, side
5 Mizzenmast, side
6 Mizzenmast, rear
7 Main topmast
8 Fore topmast
9 Mizzen topmast
10 Main topgallant mast

11 Fore topgallant mast
12 Mizzen topgallant mast
13 Main royal mast with skysail-pole
14 Fore royal mast with skysail-pole
15 Mizzen royal mast with skysail-pole
16 Cap tenon
17 Masthead hoop
18 Masthead
19 Mast battens
20 Chock
21 Bibs
22 Hounds

23 Wooldings
24 Mast hoops
25 Cheeks
26 Iron fastening bolts
27 Rubbing paunch or fish
28 Heel
29 Spanker mast
30 Spanker mast bracket
31 Topmast cap tenon
32 Cheek block
33 Masthead

34 Hounds
35 Octagonal
36 Top rope sheave
37 Heel
38 Fit hole
39 Block
40 Block sheave
41 Block hoop
42 Mizzen topmast hounds with tye sheave
43 Heel without block
44 Squared topgallant mast tenon for iron cap
45 Topgallant mast hounds with tye sheave
46 Topgallant mast heel
47 Topgallant mast fit hole
48 Truck
49 Skysail-pole
50 Royal mast stop
51 Royal mast
52 Hole for lower cap bolt

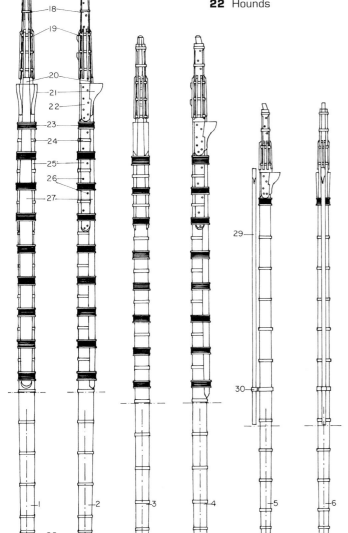

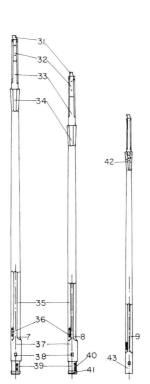

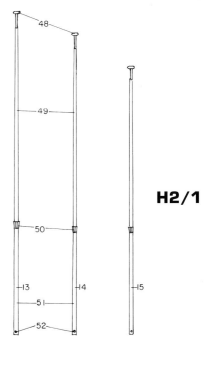

H2/1

SCALE IN FEET

0   5   10   20   30

H Masts and Spars

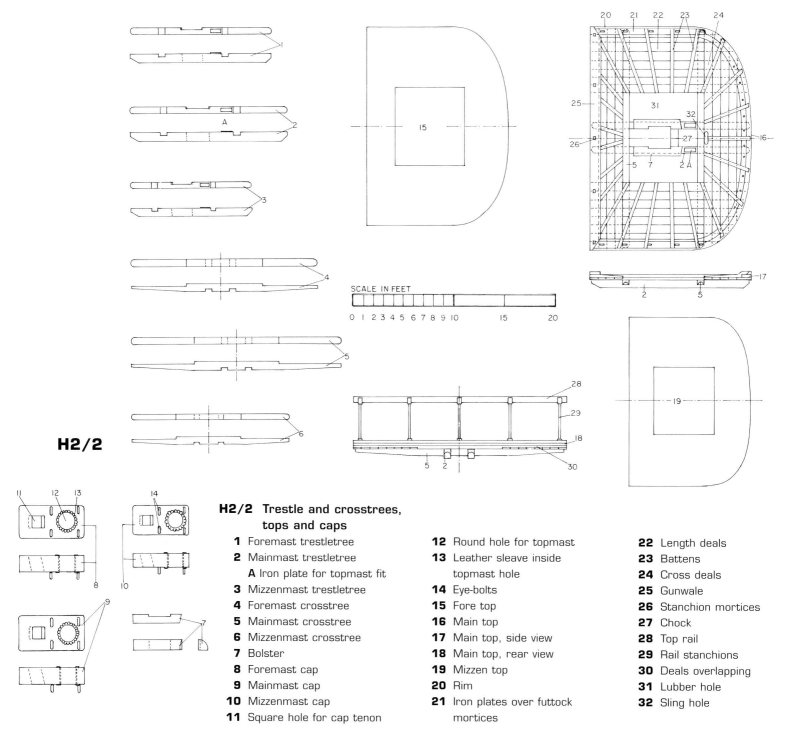

H2/2

H2/2 Trestle and crosstrees, tops and caps

1 Foremast trestletree
2 Mainmast trestletree
A Iron plate for topmast fit
3 Mizzenmast trestletree
4 Foremast crosstree
5 Mainmast crosstree
6 Mizzenmast crosstree
7 Bolster
8 Foremast cap
9 Mainmast cap
10 Mizzenmast cap
11 Square hole for cap tenon

12 Round hole for topmast
13 Leather sleave inside topmast hole
14 Eye-bolts
15 Fore top
16 Main top
17 Main top, side view
18 Main top, rear view
19 Mizzen top
20 Rim
21 Iron plates over futtock mortices

22 Length deals
23 Battens
24 Cross deals
25 Gunwale
26 Stanchion mortices
27 Chock
28 Top rail
29 Rail stanchions
30 Deals overlapping
31 Lubber hole
32 Sling hole

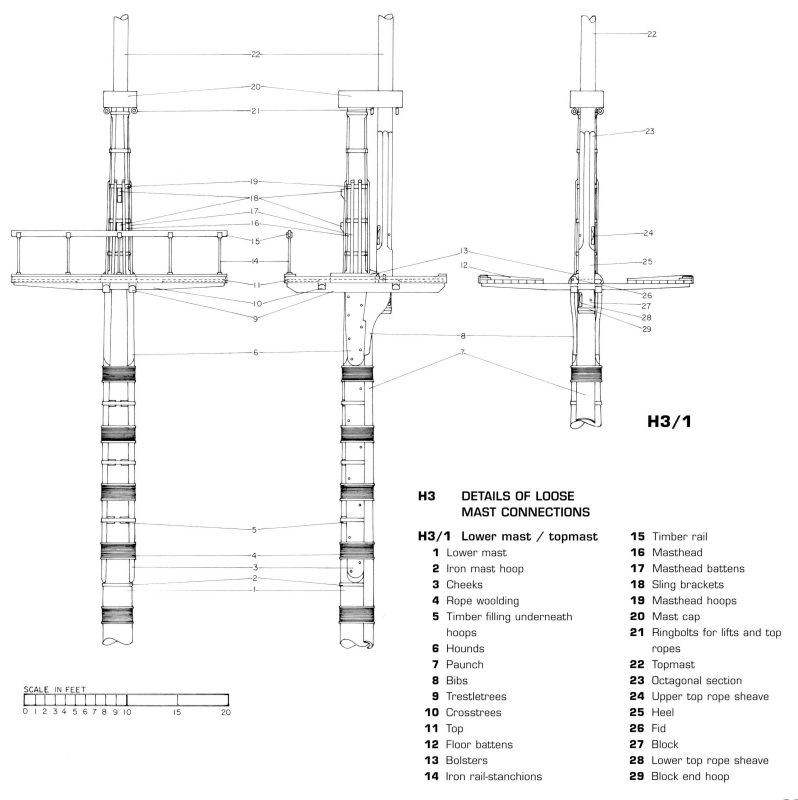

H3/1

H3   DETAILS OF LOOSE MAST CONNECTIONS

H3/1 Lower mast / topmast

1	Lower mast	**15**	Timber rail
2	Iron mast hoop	**16**	Masthead
3	Cheeks	**17**	Masthead battens
4	Rope woolding	**18**	Sling brackets
5	Timber filling underneath hoops	**19**	Masthead hoops
6	Hounds	**20**	Mast cap
7	Paunch	**21**	Ringbolts for lifts and top ropes
8	Bibs	**22**	Topmast
9	Trestletrees	**23**	Octagonal section
10	Crosstrees	**24**	Upper top rope sheave
11	Top	**25**	Heel
12	Floor battens	**26**	Fid
13	Bolsters	**27**	Block
14	Iron rail-stanchions	**28**	Lower top rope sheave
		29	Block end hoop

SCALE IN FEET

0 1 2 3 4 5 6 7 8 9 10   15   20

H Masts and Spars

H3/2 Mizzen mast / spanker mast

1 Mizzenmast
2 Spanker mast
3 Spanker mast step
4 Spanker mast hoop with boom bracket
5 Mizzenmast hoop
6 Woolding with 13 rope turns
7 Hounds
8 Bibs
9 Trestletree
10 Filling chock
11 Bolster
12 Chock
13 Mast battens

H3/3 Topmast / topgallant mast / royal mast with skysail pole

1 Topmast
2 Topmast heel
3 Topmast fid hole with fid indicated
4 Heel block
5 Lower top rope sheave
6 Block hoop
7 Upper top rope sheave
8 Octagonal part of topmast
9 Hounds
10 Crosstrees
11 Trestletrees
12 Bolster
13 Topmast head
14 Cheek block
15 Topmast cap
16 Topgallant mast
17 Topgallant mast fid hole
18 Topgallant mast heel
19 Topgallant mast top rope sheave
20 Hounds with tye sheave
21 Masthead
22 Cap tenon
23 Lower topgallant cap
24 Upper topgallant cap
25 Royal mast
26 Hounds with tye sheave
27 Skysail pole
28 Stop for royal stay
29 Truck

H3/4 Topmast trestle and crosstrees, topmast cap and topgallant mast lower and upper iron cap

1 Fore topmast trestletree with sheave for main topsail bowline in the aft end
2 Main topmast trestletree with sheave for main topgallant braces in the fore end
3 Crosstrees
4 View of main topmast trees assembled and bolsters fitted
5 Bolster
6 Cheek block
7 Topmast cap
8 Lower topgallant cap
9 Upper topgallant cap
10 Truck with two sheaves for skysail halyards

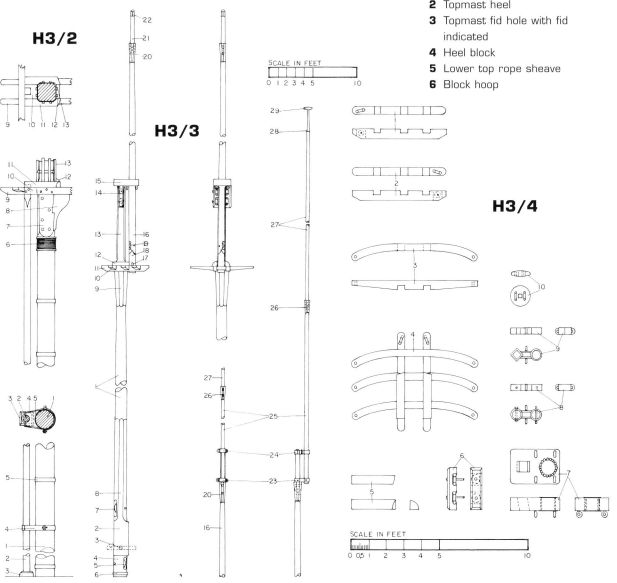

SCALE IN FEET
0 1 2 3 4 5   10

SCALE IN FEET
0 0,5 1 2 3 4 5   10

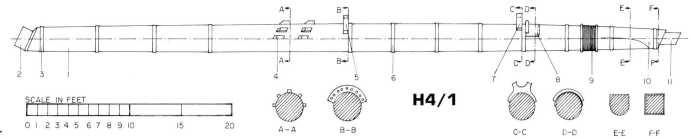

H4/1

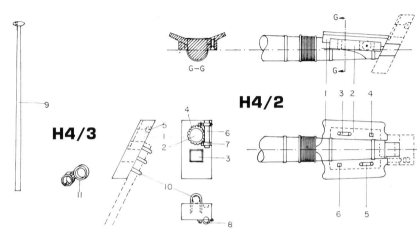

G–G

H4/6

H4/3

H4/2

H4 BOWSPRIT

H4/1 Bowsprit

1 Bowsprit
2 Heel
3 Four inner iron hoops
4 Two sets of gammoning cleats
5 Fairlead
6 Eight outer iron hoops
7 jib boom saddle
8 Spritsail sling saddle
9 Woolding
10 Bee-square
11 Cap tenon

H4/2 Bee

1 Bee
2 Bee block
3 Sheave for foremast preventer stay
4 Emergency squared hole for fore topmast preventer stay
5 Sheave for fore topmast stay
6 Emergency squared hole for fore topmast stay

H4/3 Cap

1 Bowsprit cap
2 Round hole for jib boom
3 Squared hole for cap tenon
4 Leather sleeve in round hole
5 Mortice for flying jib boom tenon
6 Half-round cut-out for jack staff
7 Iron bracket for jack staff heel
8 Hinged iron bracket for jack staff
9 Jack staff
10 Cramps for dolphin striker
11 jib boom cap

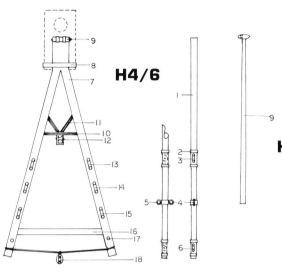

H4/6

H4/4

H4/5

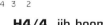

H4/4 jib boom

1 jib boom
2 Octagonal heel
3 Sheave for jib boom outhaul
4 Hole for jib boom lashing
5 Sheave for standing jib stay
6 Stop
7 Octagonal cap tenon

H4/5 Flying jib boom

1 Flying jib boom
2 Octagonal heel
3 Hole for lashing
4 Flying jib boom tenon
5 Sheave for flying jib stay
6 Stop

H4/6 Dolphin strikers

1 Single dolphin striker in use until around 1803
2 Iron hoops
3 Sheave for standing jib stay
4 Eyebolt for double martingale forward to jib and flying jib stops
5 Eyebolt for double martingale rearward to bow
6 Sheave for flying jib stay
7 Double dolphin striker after approx. 1803
8 Iron strap around the cap
9 Hinge
10 Iron bar

11 Iron braces
12 Iron roller for martingale stay
13 Upper sheave for martingale guy
14 Lower sheave for martingale guy
15 Upper sheave for flying jib martingale guy
16 Timber spreader
17 Hole for lower flying jib martingale guy
18 Block for traveller guy

H Masts and Spars

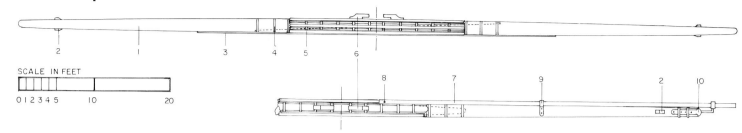

SCALE IN FEET

0 1 2 3 4 5   10   20

H5   SPARS

H5/1 Yards

1 Fore and main lower yard
2 Stop cleat
3 Two inch thick back fish
4 Iron hoops for larger 'built' yards
5 Battens for the octagonal quarter except for front and rear

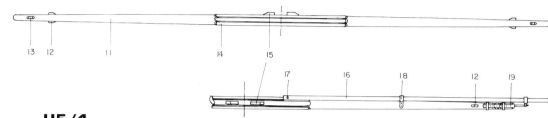

H5/1

6 Sling cleats
7 Topsail studdingsail boom
8 Hole for lashing
9 Quarter boom iron
10 Outer boom iron
11 Fore & main topsail yard
12 Stop cleats
13 Reef-tackle sheave
14 Battens for octagonal quarter, except the front
15 Tye cleats
16 Topgallant studdingsail boom
17 Hole for lashing
18 Quarter boom iron
19 Outer boom iron
20 Fore & main topgallant sail yard

21 Stop cleats
22 Iron ferule
23 Eyebolt for jewel block
24 Tye cleats
25 Royal yard
26 Stop cleats
27 Iron ferule
28 Tye cleats
29 Crossjack yard
30 Stop cleats
31 Yard arm (outside stop cleats)
32 Iron ferule

33 Sixteen-sided inner quarter
34 Sling cleats
35 Mizzen topsail yard
36 Stop cleats
37 Reef-tackle sheave
38 Iron ferule
39 Octagonal quarter
40 Tye cleats
41 Mizzen topgallant sail yard
42 Stop cleats
43 Iron ferule
44 Tye cleats

H5/2 Gaff, spanker boom, studdingsail yards and swinging lower main studdingsail boom

1 Crossjack yard (not to scale)
2 Ferule
3 Yardarm stop cleats
4 Sling cleats
5 Gaff
6 Eyebolt for ensign block
7 Ferule
8 Peak stop cleats

9 Brail stop cleats

10 Yaw tongue hoops

11 Tongue

12 Eyebolt for gaff halyard

13 Gaff yaws

14 Spanker boom

15 End cap

16 Spanker sheet sheave

17 Topping lift stop cleat

18 Boom sheet eyebolts

19 Eyebolt for spanker
sheet tackle

20 Yaw tongue hoops

21 Eyebolt for spanker tack

22 Boom yaws

23 Ringtail boomkin

24 Stop

25 Hole for lashing

26 Bumpkin or boomkin

27 Eyebolts for bumpkin shrouds

28 Swinging lower studdingsail
boom

29 Ferule

30 Stop cleat for tack block

31 Stop cleats for martingale
and topping lift

32 Gooseneck

33 Lower studdingsail yard

34 Ferule

35 Yardarm stop cleats

36 Halyard stop cleats

37 Topsail studdingsail yard

38 Ferule

39 Yardarm stop cleats

40 Halyard stop cleats

41 Topgallant studdingsail yard

42 Ferule

43 Yardarm stop cleats

44 Halyard stop cleats

H5/3   Yard details

1 Lower yard

2 Boom quarter iron with hinge

3 Yardarm stop cleats

4 Outer boom iron

5 Outer boom iron with iron
roller as seen from the end

6 Topsail yardarm

7 Reef-tackle sheave

8 Outer boom iron with
stop

9 45° squared arm and

10 Threaded end

11 Loose boom iron part

12 Nut

13 Inner gaff end

14 Tongue

15 Iron hoops

16 Bolts

17 Eyebolt for halyard

18 Eyebolt for downhaul

19 Leather inlay

20 Yaws half-round in 40°

21 Hole for single parrel

22 Inner boom end

23 Iron hoops

24 Tongue

25 Bolts

26 Eyebolt for spanker tack

27 Leather inlay

28 Yaws half-round

29 Hole for single parrel

30 Outer boom end

31 Spanker sheet sheave

32 Boom cap for ringtail boomkin

33 Fitted swinging lower
studdingsail boom

34 Channel

35 Turning iron

36 Resting iron

H5/2

SCALE IN FEET

0 1 2 3 4 5   10   20

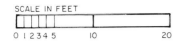

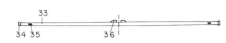

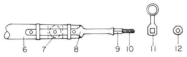

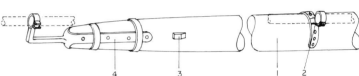

H5/3

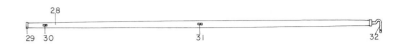

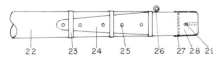

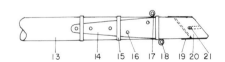

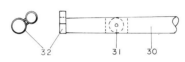

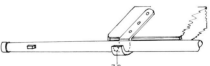

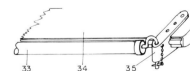

J Standing Rigging

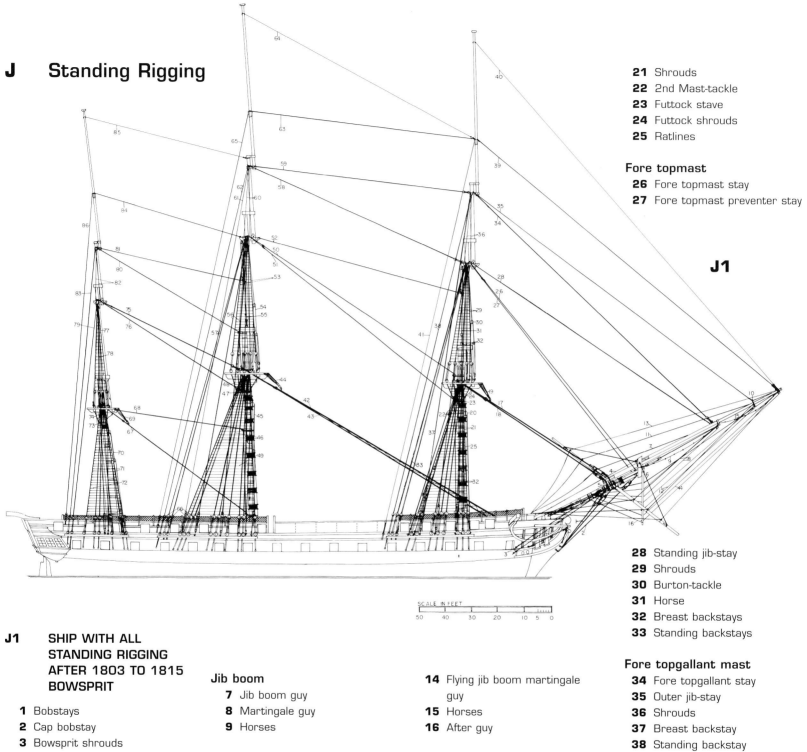

J1

J1 SHIP WITH ALL STANDING RIGGING AFTER 1803 TO 1815

BOWSPRIT
1 Bobstays
2 Cap bobstay
3 Bowsprit shrouds
4 Bowsprit horse

Spritsail yard
5 Slings
6 Standing lifts

Jib boom
7 Jib boom guy
8 Martingale guy
9 Horses

Flying jib boom
10 Traveller
11 Traveller guy
12 Traveller martingale guy
13 Flying jib boom guy

14 Flying jib boom martingale guy
15 Horses
16 After guy

Fore mast
17 Fore stay
18 Fore preventer stay
19 Crows-feet
20 Mast-tackle

21 Shrouds
22 2nd Mast-tackle
23 Futtock stave
24 Futtock shrouds
25 Ratlines

Fore topmast
26 Fore topmast stay
27 Fore topmast preventer stay

28 Standing jib-stay
29 Shrouds
30 Burton-tackle
31 Horse
32 Breast backstays
33 Standing backstays

Fore topgallant mast
34 Fore topgallant stay
35 Outer jib-stay
36 Shrouds
37 Breast backstay
38 Standing backstay

Fore royal mast
39 Flying jib-stay
40 Royal stay
41 Standing backstay

SCALE IN FEET
50 40 30 20 10 5 0

Mainmast

42 Mainstay
43 Main preventer stay
44 Crows-feet
45 Mast-tackle
46 Shrouds
47 Futtock stave
48 Futtock shrouds
49 Ratlines

Main topmast

50 Main topmast stay
51 Main topmast preventer stay
52 Middle staysail stay
53 Shrouds
54 Burton-tackle
55 Horse
56 Breast backstays
57 Standing backstays

Main topgallant mast

58 Main topgallant stay
59 Main topgallant staysail stay
60 Shrouds
61 Breast backstay
62 Standing backstay

Main royal mast

63 Royal staysail stay
64 Royal stay
65 Standing backstay
66 Dead eyes

Mizzen mast

67 Mizzen stay
68 Mizzen staysail stay
69 Crows-feet
70 Mast-tackle
71 Shrouds
72 Ratlines
73 Futtock stave
74 Futtock shrouds

Mizzen topmast

75 Mizzen topmast stay
76 Mizzen topmast staysail stay
77 Shrouds

78 Breast backstay
79 Standing backstay

Mizzen topgallant mast

80 Mizzen topgallant stay
81 Mizzen topgallant staysail stay
82 Shrouds
83 Standing backstay
84 Standing backstay

Mizzen royal mast

85 Mizzen royal staysail stay
86 Mizzen royal stay

J1/1 Shrouds and backstays fitted to channel

1 Mast tackle with pendant, runner and fall
2 First shroud pair (on starboard) served over for protection
3 Ratlines
4 First topmast breast backstay
5 Shroud-bound dead eye

6 Long stropped lower mast tackle fall block wit hook
7 Ironbound dead eye
8 Iron fitting for the swinging boom gooseneck
9 Channel
10 Chain
11 Eyebolt on channel
12 Iron fitting for resting the swinging boom
13 Royalmast standing backstay
14 Topgallant mast standing backstay
15 Topmast standing backstay
16 Topgallant mast breast backstay
17 Topsail yard tie tackle fall
18 Shrouds
19 Second topmast breast backstay
20 Lanyards

J1/2 Shrouds and stays fitted on masthead

1 Lower stay
2 Lower preventer stay
3 Mouse and eye (pointed over)

4 Snaking
5 Mast tackle pendant with thimble
6 Mast tackle runner block hooked into the thimble
7 Stropped in upper mast tackle fall block
8 Shrouds
9 Ratlines
10 Futtock stave
11 Catharpin
12 Seized eye of a shroud pair

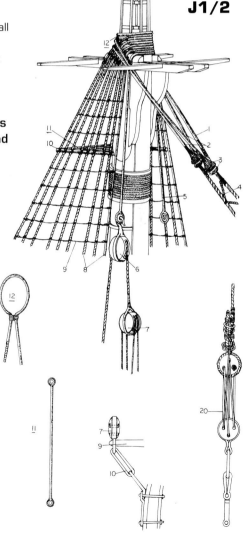

J1/2

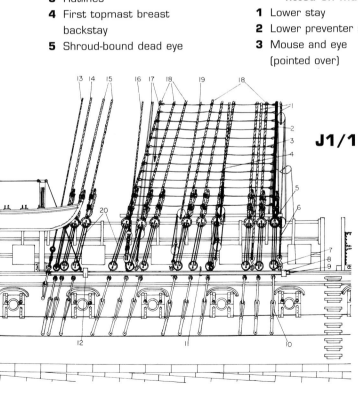

J1/1

101

J Standing Rigging

J1/3 Mast top with futtock shrouds

1 Futtock dead eye
2 Futtock shroud
3 Ratline
4 Catharpin
5 Futtock stave
6 Futtock shroud lashed to mast shroud
7 Shroud
8 Futtock dead eye with hooked in futtock shroud

J1/4 Topmast head with topmast and topgallant shrouds

1 Topgallant masthead
2 Capelage
3 Topgallant shrouds
4 Topgallant backstays
5 Topgallant shrouds
6 opmast-head
7 Topmast shrouds
8 Sisterblock
9 Futtock stave
10 Pendant for burton tackle
11 Burton tackle
12 Thimble spliced into lower end of topgallant shrouds to be lashed to futtock dead eyes

J1/5 Royal mast and skysail pole

1 Skysail pole
2 Grommet
3 Royal mast hounds
4 Royal backstay
5 Topgallant masthead

J2 STAYS

J2/1 Lower stays and top details

1 Crowfeet
2 Euphroe with tackle
3 Top rim holes for crowfeet
4 Preventer stay
5 Snaking
6 Lower stay
7 Turned in heart
8 Lanyard with lashing
9 Forelocked eyebolt for lower heart
10 Collar
11 Mouse
12 Eye
13 Stay
14 Worming
15 Seizings
16 Heart

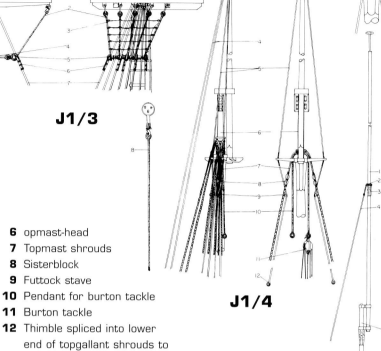

J1/3

J1/4

J1/5

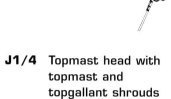

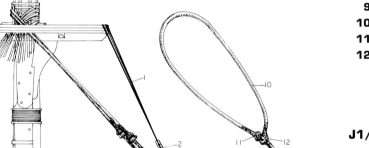

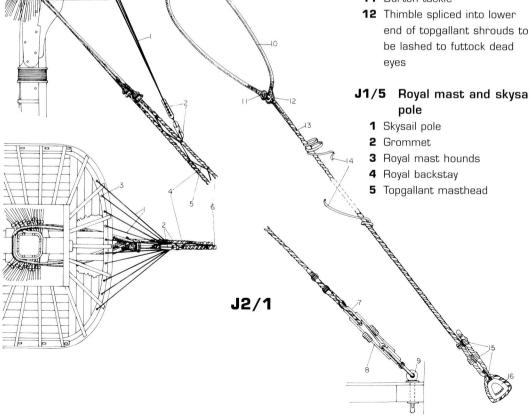

J2/1

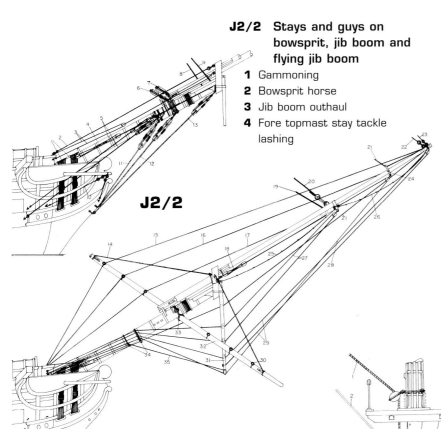

J2/2 Stays and guys on bowsprit, jib boom and flying jib boom

1 Gammoning
2 Bowsprit horse
3 Jib boom outhaul
4 Fore topmast stay tackle lashing

5 Bowsprit shrouds
6 Fore preventer stay heart with collar
7 Fore stay with collar
8 Fore topmast preventer stay
9 Fore topmast stay
10 Bumpkin shrouds
11 Inner bobstay
12 Second bobstay
13 Cap bobstay
14 Standing lift
15 Flying jib boom guy
16 Traveller guy
17 Jib boom guy
18 Standing jib stay lashing
19 Standing jib stay

20 Fore topgallant stay
21 Outer jib stay
22 Flying jib stay
23 Royal stay
24 Traveller
25 Jib boom horse
26 Flying jib boom horse
27 Martingale guy
28 Traveller martingale stay
29 Flying jib boom martingale guy
30 Guide thimbles
31 Dolphin striker (Martingale)
32 Spritsail yard
33 Sling
34 Guide thimbles
35 After martingale guy

J 2/3 Main and mizzen stays as fitted at their lower ends

1 Main topmast stay to foremast
2 Main topmast preventer stay to foremast
3 Mizzen stay to mainmast
4 Mizzen staysail stay to mainmast
5 Mizzen topmast preventer stay to mainmast
6 Mizzen topmast stay to mainmast
7 Mizzen topgallant stay to mainmast
8 Mizzen middle staysail stay to main topmast

9 Mizzen royal stay to main topmast
10 Mizzen skysail pole stay to main topgallant mast
11 Main middle staysail stay to fore topmast
12 Main topgallant stay to fore topmast
13 Main topgallant staysail stay to fore topgallant mast
14 Main royal stay to fore royal mast
15 Main skysail pole stay to fore royal mast

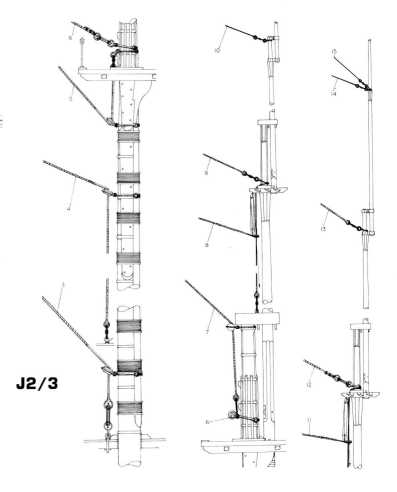

J2/3

K Running Rigging

K1

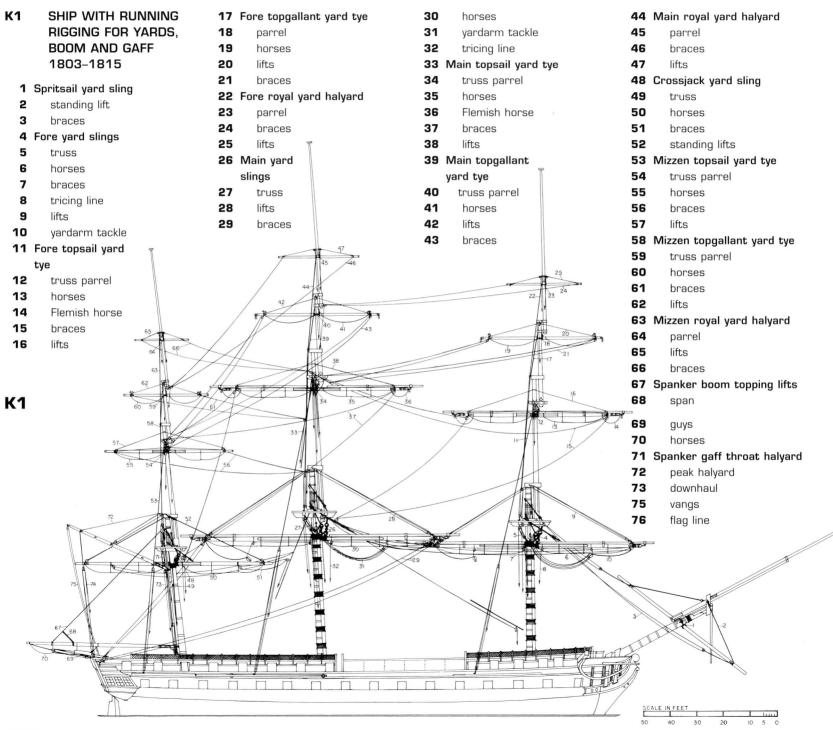

SCALE IN FEET

50   40   30   20   10 5 0

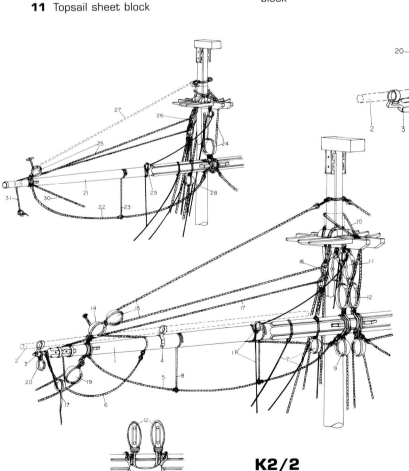

K2 YARD RIGGING DETAILS

K2/1 Main and fore yard

1 Yard
2 Topsail studdingsail boom
3 Outer boom iron
4 Inner boom iron
5 Sling
6 Preventer sling
7 Sling lanyard lashing
8 Horses
9 Stirrups
10 Lifts
11 Topsail sheet block

12 Quarter block
13 Slab line with block
14 Truss
15 Yardarm tackle pendant
16 Tricing line with block
17 Clew garnet and block
18 Buntlines
19 Leech line
20 Dog & bitch connected brace block

K2/1

K2/2 Topsail yards

1 Main and fore topsail yard
2 Topgallant studdingsail boom
3 Outer boom iron
4 Inner boom iron
5 Horses
6 Flemish horse
7 Clew line and block
8 Stirrups
9 Quarter block
10 Double tye
11 Upper tye blocks
12 Yard tye blocks
13 Leather clad truss parrel
14 Topgallant sail sheet block
15 Lift with lift block
16 Sisterblock
17 Reef tackle with tail block

18 Buntline with block
19 Brace block (dog & bitch)
20 Jewel block
21 Mizzen topsail yard
22 Horses
23 Stirrups
24 Single tye block and tye
25 Topgallant sheet and lift block
26 Sisterblock
27 Topgallant sheet locked in as lift when topgallant sail is furled
28 Clew line block
29 Buntline block
30 Braces
31 Reef tackle

K2/2

K Running Rigging

K2/3 Crossjack yard, main and fore topgallant yards, royal yards

1 Crossjack yard
2 Horses
3 Stirrups
4 Braces
5 Quarter block
6 Sling
7 Topsail sheet block
8 Standing lifts
9 Truss
10 Main and fore topgallant yards
11 Horses
12 Clew block
13 Buntline and block
14 Tye
15 Royal sheet and block
16 Braces
17 Jewel block
18 Royal yards and mizzen topgallant yard
19 Tye
20 Lifts
21 Braces

K2/3

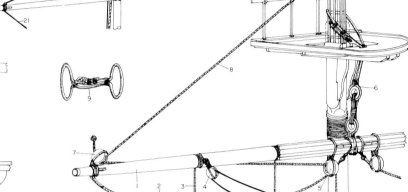

K2/5

K2/4

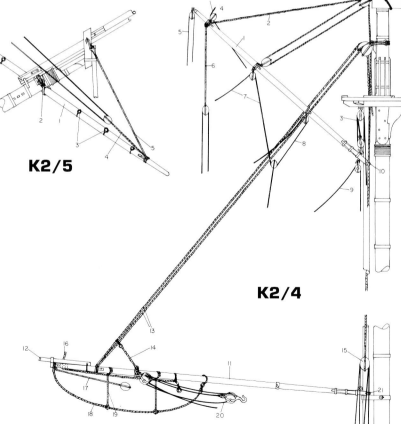

K2/4 Spanker gaff and boom

1 Gaff
2 Peak halyard
3 Throat halyard
4 Gaff topsail sheet
5 Flag line
6 Vangs
7 Peak brails
8 Middle brails
9 Foot brails
10 Single parrel
11 Boom
12 Ringtail bumpkin
13 Topping lifts
14 Span
15 Topping lift tackle
16 Spanker sheet
17 Boom guy
18 Horse
19 Stirrups
20 Boom sheet
21 Single parrel

K2/5 Spritsail yard

1 Spritsail yard
2 Sling
3 Guide thimbles
4 Braces
5 Standing lifts

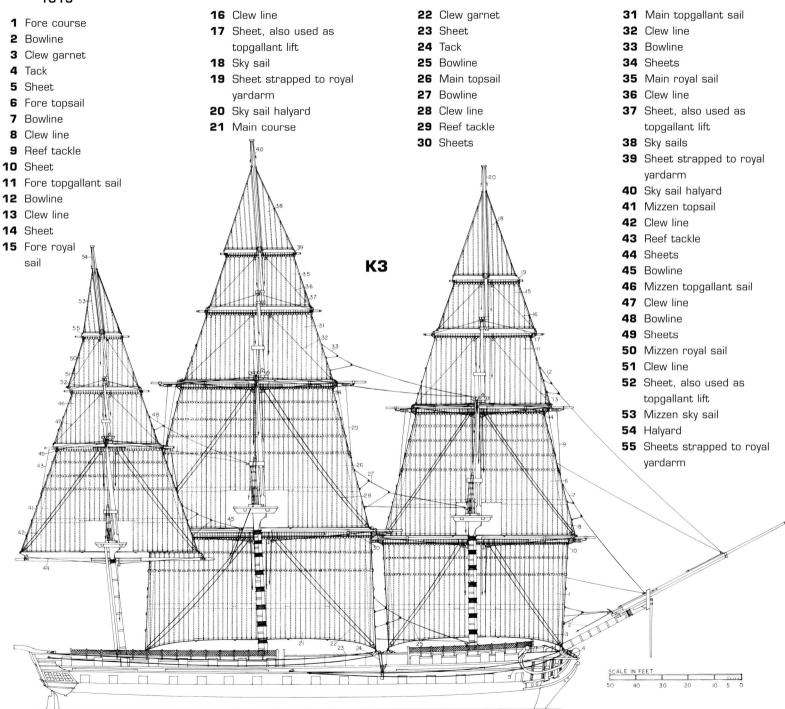

K3 SHIP WITH SQUARE-
SAILS AS RIGGED
BETWEEN 1803 AND
1815

1 Fore course
2 Bowline
3 Clew garnet
4 Tack
5 Sheet
6 Fore topsail
7 Bowline
8 Clew line
9 Reef tackle
10 Sheet
11 Fore topgallant sail
12 Bowline
13 Clew line
14 Sheet
15 Fore royal
sail

16 Clew line
17 Sheet, also used as
topgallant lift
18 Sky sail
19 Sheet strapped to royal
yardarm
20 Sky sail halyard
21 Main course

22 Clew garnet
23 Sheet
24 Tack
25 Bowline
26 Main topsail
27 Bowline
28 Clew line
29 Reef tackle
30 Sheets

31 Main topgallant sail
32 Clew line
33 Bowline
34 Sheets
35 Main royal sail
36 Clew line
37 Sheet, also used as
topgallant lift
38 Sky sails
39 Sheet strapped to royal
yardarm
40 Sky sail halyard
41 Mizzen topsail
42 Clew line
43 Reef tackle
44 Sheets
45 Bowline
46 Mizzen topgallant sail
47 Clew line
48 Bowline
49 Sheets
50 Mizzen royal sail
51 Clew line
52 Sheet, also used as
topgallant lift
53 Mizzen sky sail
54 Halyard
55 Sheets strapped to royal
yardarm

K3

SCALE IN FEET
50 40 30 20 10 5 0

K Running Rigging

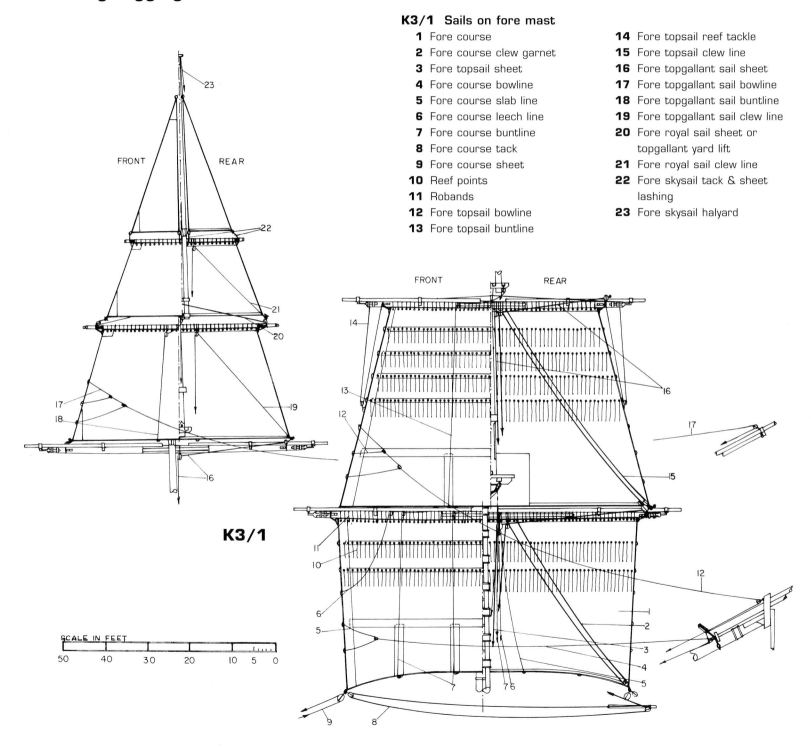

K3/1

SCALE IN FEET
50   40   30   20   10 5   0

K3/2 Sails on main mast

1 Main course
2 Main course clew garnet
3 Main topsail sheet
4 Main course slab line
5 Main course buntline
6 Main course leech line
7 Main course bowline
8 Main course sheet
9 Main course tack
10 Reef points
11 Robands
12 Main topsail bowline
13 Main topsail clew line
14 Main topsail reef tackle
15 Main topsail buntline
16 Main topgallant sail sheet
17 Main topgallant sail clew line
18 Main topgallant sail bowline
19 Main topgallant sail buntline

K3/3 Main course bowlines on fore mast (no scale)

1 Stropped double block lashed to rear of lower fore mast
2 Fore mast
3 Bowlines belayed to jeer bits
4 Starboard bowline
5 Portside bowline

K3/4 Course clew with block attachment (no scale)

1 Clew
2 Bolt rope (leech rope)
3 Serviced part of leech rope
4 Clew garnet block
5 Tack block seized to clew
6 Sheet block
7 Service part of foot rope

K3/3

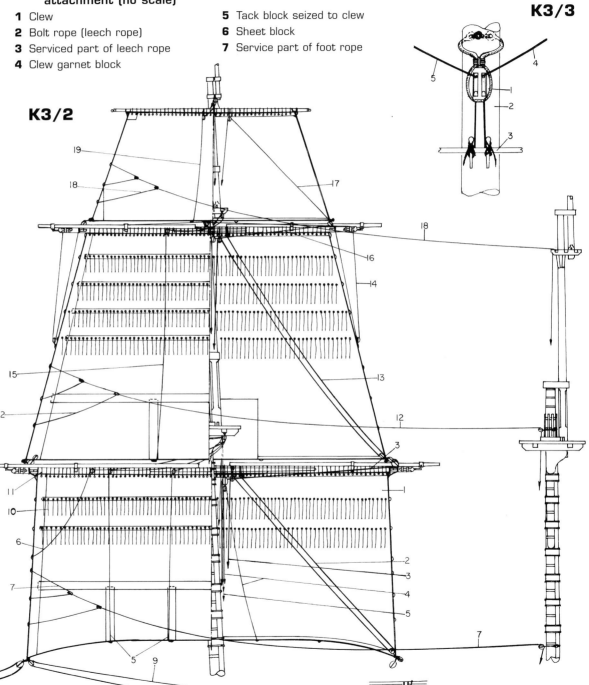

K3/2

K3/4

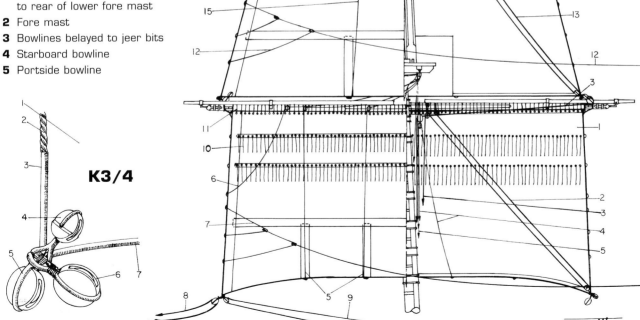

109

K   Running Rigging

K3/5 Square sails of mizzen mast

1 Mizzen topsail
2 Mizzen topsail sheet
3 Mizzen topsail clew line
4 Reef points
5 Mizzen topsail reef tackle
6 Mizzen topsail bowline
7 Mizzen topsail buntline
8 Robands
9 Mizzen topgallant sail sheet
10 Mizzen topgallant sail bowline
11 Mizzen topgallant sail buntline
12 Mizzen topgallant sail clew line
13 Mizzen royal sail sheet or topgallant yard lift
14 Mizzen royal sail clew line
15 Mizzen skysail tack and sheet lashings
16 Mizzen skysail halyards

K3/6 Sail details

1 Skysail sheet and tack lashed to royal yard with sheet lashing still incomplete
2 Royal and topgallant sail clew with a toggled sheet
3 Reef points, the shorter length at the front of sail
4 Topsail sheet with double wall knot

K4   STUDDING SAILS

1 Lower main studding sail, lower fore studding sail is without reef points
2 Reef points
3 Swinging studding sail boom for main channel
4 Tack
5 Fore guy
6 Swinging boom martingale
7 Swinging boom topping lift
8 Aft guy
9 Fore sheet
10 Aft sheet
11 Lower studding sail yard
12 Topmast studding sail boom
13 Lower studding sail inner halyard
14 Topmast studding sail boom brace
15 Lower studding sail halyard

K3/5

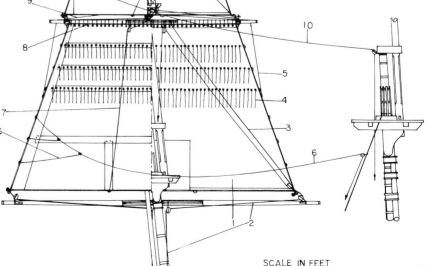

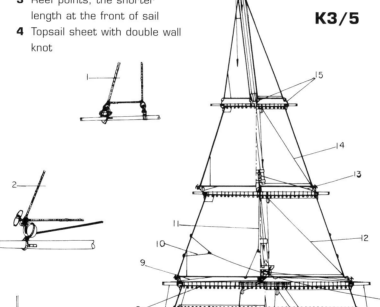

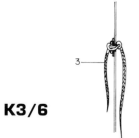

K3/6

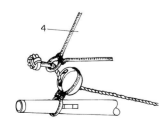

SCALE IN FEET
50   40   30   20   10 5 0

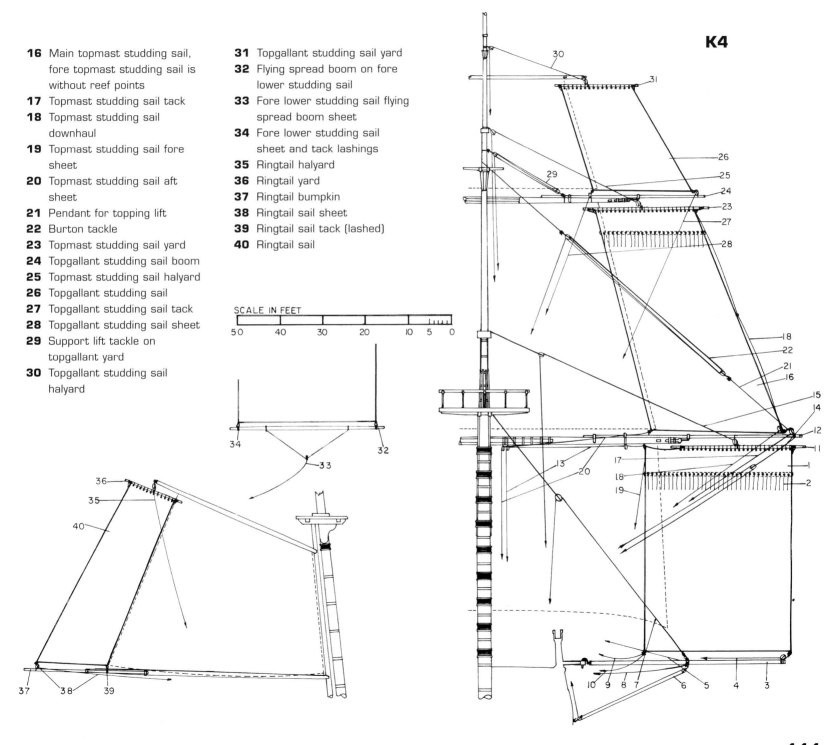

16 Main topmast studding sail,
fore topmast studding sail is
without reef points

17 Topmast studding sail tack

18 Topmast studding sail
downhaul

19 Topmast studding sail fore
sheet

20 Topmast studding sail aft
sheet

21 Pendant for topping lift

22 Burton tackle

23 Topmast studding sail yard

24 Topgallant studding sail boom

25 Topmast studding sail halyard

26 Topgallant studding sail

27 Topgallant studding sail tack

28 Topgallant studding sail sheet

29 Support lift tackle on
topgallant yard

30 Topgallant studding sail
halyard

31 Topgallant studding sail yard

32 Flying spread boom on fore
lower studding sail

33 Fore lower studding sail flying
spread boom sheet

34 Fore lower studding sail
sheet and tack lashings

35 Ringtail halyard

36 Ringtail yard

37 Ringtail bumpkin

38 Ringtail sail sheet

39 Ringtail sail tack (lashed)

40 Ringtail sail

K4

SCALE IN FEET

50 40 30 20 10 5 0

K Running Rigging

K5 SHIP WITH FORE AND AFT SAILS AS RIGGED BETWEEN 1803 AND 1815

1 Flying jib
2 Halyard
3 Sheets
4 Outer jib stay
5 Outer jib
6 Halyard
7 Downhaul
8 Hanks
9 Traveller
10 Sheets
11 Jib
12 Halyard
13 Downhaul
14 Sheets
15 Fore topmast staysail
16 Halyard
17 Downhaul
18 Hanks
19 Sheets
20 Main topmast staysail
21 Halyard

22 Lifter and Downhaul
23 Hanks
24 Nock
25 Tack
26 Sheets
27 Middle staysail
28 Halyard
29 Downhaul
30 Nock
31 Sheets
32 Tack
33 Main topgallant staysail
34 Halyard
35 Downhaul
36 Nock
37 Tack
38 Sheets
39 Main royal staysail
40 Halyard
41 Hanks

42 Downhaul
43 Nock
44 Tack
45 Sheets
46 Mizzen staysail
47 Halyard
48 Downhaul
49 Lifter
50 Nock
51 Tack
52 Sheet
53 Mizzen topmast staysail
54 Halyard
55 Downhaul
56 Hanks
57 Nock
58 Tack
59 Sheets

60 Mizzen topgallant staysail
61 Halyard
62 Downhaul
63 Nock
64 Tack
65 Sheets
66 Stay lifter
67 Mizzen royal staysail
68 Halyard
69 Downhaul
70 Hanks
71 Nock
72 Tack
73 Sheets
74 Spanker
75 Peak brails
76 Middle brails
77 Throat brails
78 Gaff lacing
79 Nock
80 Tack
81 Sheet
82 Gaff topsail
83 Halyard
84 Tack
85 Sheet

K5

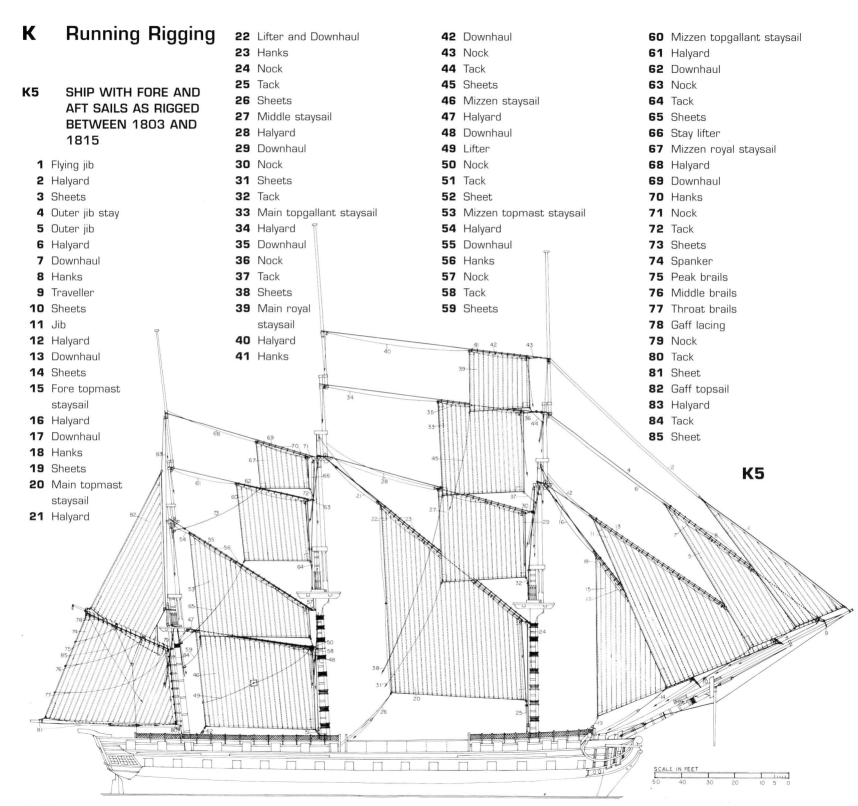

K5/1 Head sails

1 Flying jib
2 Nock lashed to flying jib boom stop
3 Single sheet (doubled)
4 Halyard
5 Outer jib
6 Outer jib stay
7 Traveller on flying jib boom
8 Downhaul
9 Hanks
10 Halyard
11 Halyard tackle (hooked to top)
12 Double sheet (doubled)

13 Nock lashed to traveller
14 Jib
15 Jib stay
16 Downhaul
17 Hanks
18 Nock downhaul block
19 Nock lashed to flying jib boom
20 Double sheet (doubled)
21 Halyard leading through portside cheek block
22 Fore topmast staysail
23 Fore topmast stay
24 Downhaul
25 Hanks

26 Nock lashed to lower stay
27 Double sheet (doubled with short pendants)
28 Grommet
29 Halyard leading through starboard cheek block
30 Cheek block

K5/1

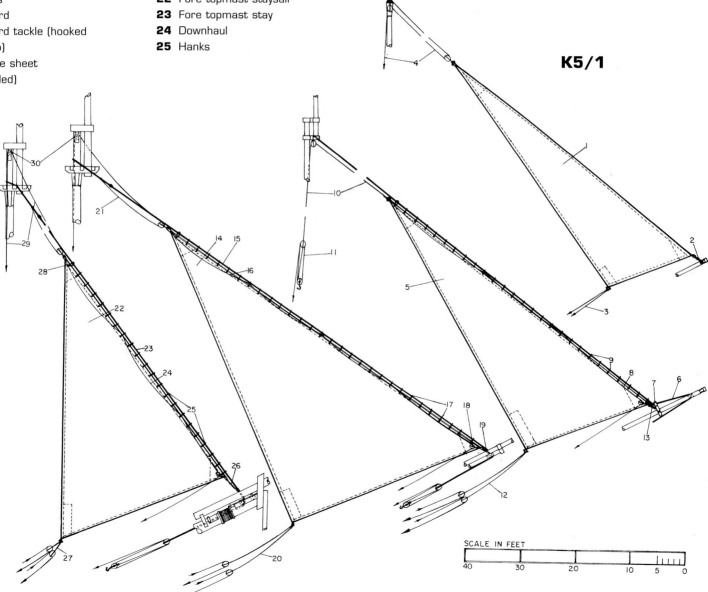

SCALE IN FEET

40   30   20   10   5   0

K Running Rigging

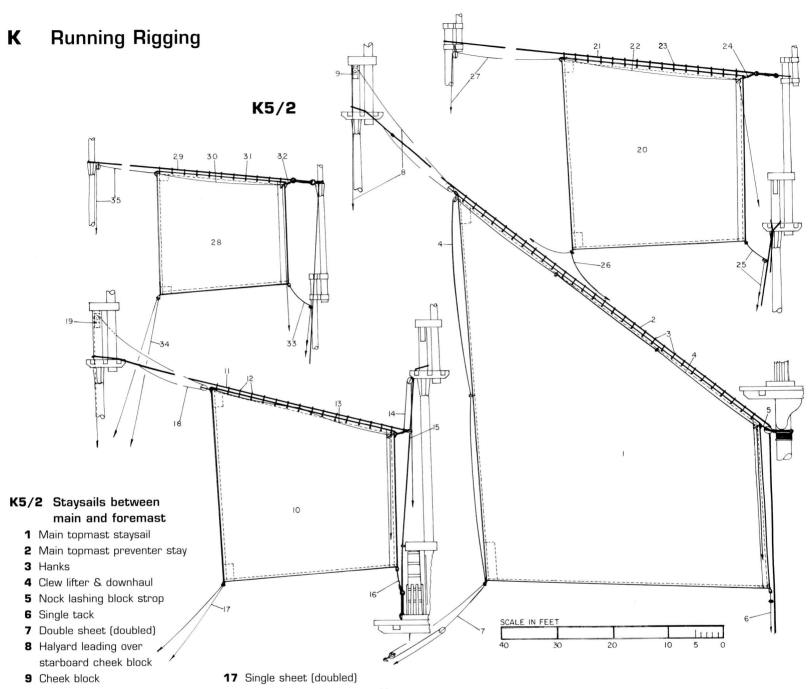

K5/2

K5/2 Staysails between main and foremast

1 Main topmast staysail
2 Main topmast preventer stay
3 Hanks
4 Clew lifter & downhaul
5 Nock lashing block strop
6 Single tack
7 Double sheet (doubled)
8 Halyard leading over starboard cheek block
9 Cheek block
10 Main middle staysail
11 Middle staysail stay
12 Hanks
13 Downhaul
14 Middle staysail tricing line
15 Jack stay
16 Tack lashed to horse

17 Single sheet (doubled)
18 Halyard leading over portside cheek block
19 Portside cheek block
20 Main topgallant staysail
21 Main topgallant stay
22 Downhaul
23 Hank

24 Nock lashing
25 Tack (doubled)
26 Single sheet (doubled)
27 Halyard
28 Main royal staysail
29 Main royal stay

30 Downhaul
31 Hank
32 Nock seizing
33 Tack (doubled)
34 Single sheet (doubled)
35 Halyard

SCALE IN FEET
40   30   20   10 5 0

K5/3 Staysails between mizzen and mainmast

1 Mizzen staysail
2 Mizzen staysail stay
3 Downhaul
4 Hank
5 Nock seized to lead block strop
6 Double block
7 Brails
8 Tack lashing
9 Sheet
10 Halyard
11 Mizzen topmast staysail

12 Mizzen topmast staysail stay
13 Hank
14 Downhaul
15 Nock seized to lead block
16 Tack (doubled)
17 Single sheet (doubled)
18 Halyard
19 Mizzen topgallant staysail
20 Mizzen topgallant staysail stay
21 Downhaul

22 Hank
23 Nock seized to stay
24 Tricing line
25 Jack stay
26 Tack
27 Single sheet (doubled)

28 Halyard
29 Mizzen royal staysail
30 Mizzen royal staysail stay
31 Downhaul
32 Hank
33 Nock (seized)
34 Tack (doubled)
35 Single sheet (doubled)
36 Halyard

K5/3

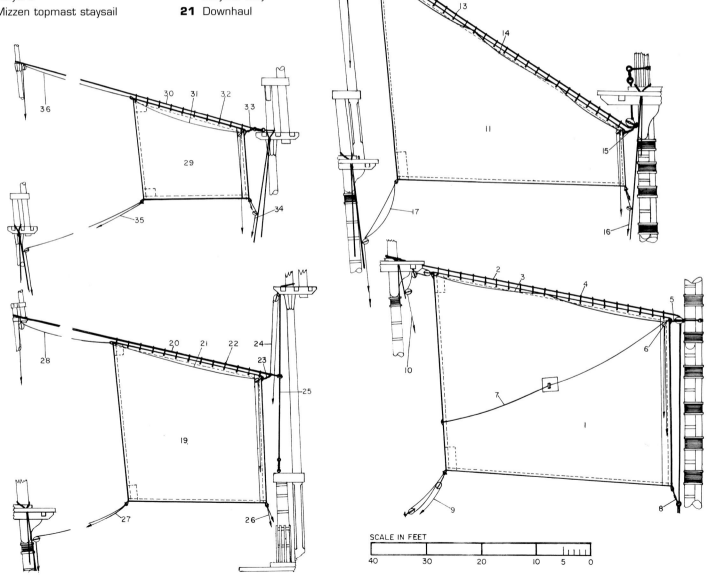

SCALE IN FEET

40   30   20   10 5 0

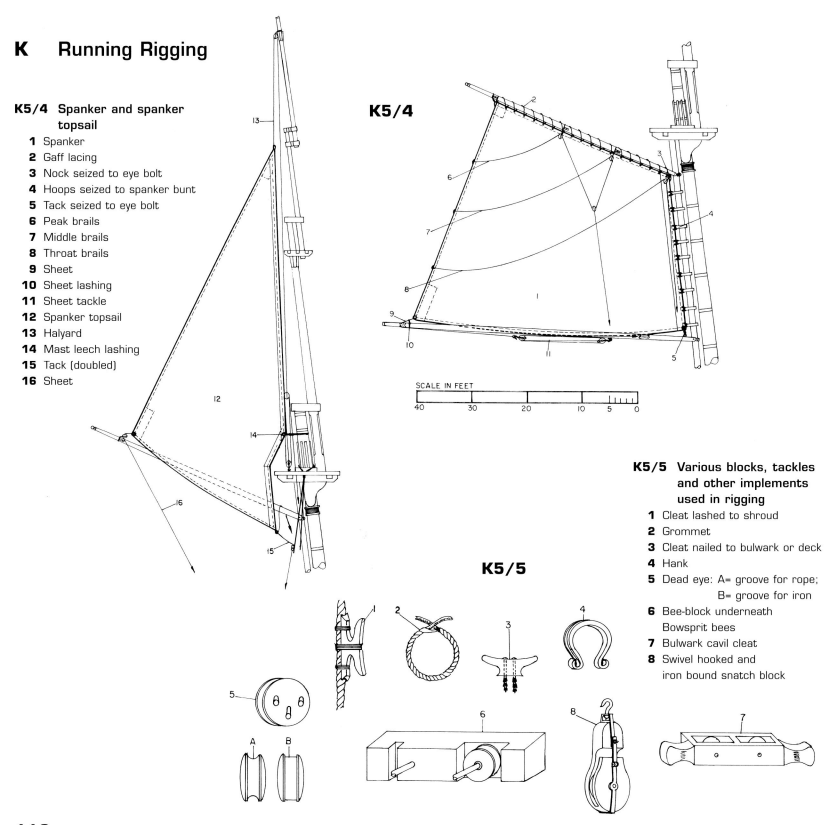

K Running Rigging

K5/4 Spanker and spanker topsail

1 Spanker
2 Gaff lacing
3 Nock seized to eye bolt
4 Hoops seized to spanker bunt
5 Tack seized to eye bolt
6 Peak brails
7 Middle brails
8 Throat brails
9 Sheet
10 Sheet lashing
11 Sheet tackle
12 Spanker topsail
13 Halyard
14 Mast leech lashing
15 Tack (doubled)
16 Sheet

K5/4

SCALE IN FEET
40   30   20   10   5   0

K5/5 Various blocks, tackles and other implements used in rigging

1 Cleat lashed to shroud
2 Grommet
3 Cleat nailed to bulwark or deck
4 Hank
5 Dead eye: A= groove for rope;
              B= groove for iron
6 Bee-block underneath Bowsprit bees
7 Bulwark cavil cleat
8 Swivel hooked and iron bound snatch block

K5/5

116

K5/5

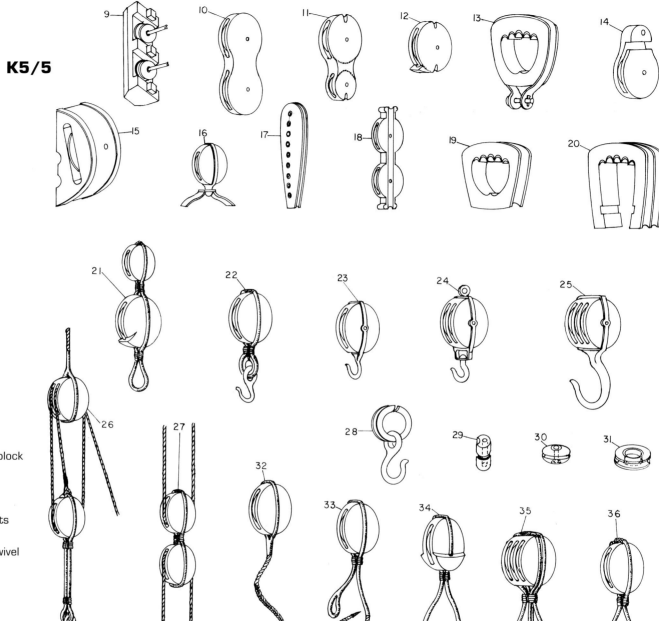

9 Cheek block
10 Leg and fall block
11 Fiddle or long-tackle-block
12 Shoulder block
13 Iron bound heart
14 Normal snatch block
15 D-block for leading lifts inboard
16 Monkey block with swivel
17 Euphroe
18 Sisterblock
19 Heart
20 Open heart
21 Shoulder block and lift block stropped together
22 Stropped single block with hook
23 Iron bound single block with hook
24 Iron bound double block with swivel hook
25 Treble sheaved cat-block

26 A tackle with a long-stropped block fitted with thimble and hook
27 Two single blocks stropped to form a leg and fall block
28 Hook and thimble

29 Fairlead truck
30 Wooden thimble
31 Different wooden thimble
32 Single block stropped with tail
33 Single block stropped with short and long leg

34 Clew garnet block
35 Double stropped treble block
36 Stropped single block with eye

K Running Rigging

K6 Belay positions

1 Bowsprit
2 Flying jib boom guy
3 Traveller guy
4 Jib boom guy
5 Jib boom outhaul (portside)
6 Flying jib boom martingale stay
7 Outer jib stay
8 Traveller martingale stay (starboard)
9 Martingale guy
10 Fore topmast stay
11 Fore topmast preventer stay
12 After martingale guy
13 Spritsail yard brace

Flying jib

13 Halyard (portside)
14 Sheet

Outer jib

15 Halyard (starboard)
16 Downhaul (starboard)
17 Sheet

Jib

18 Halyard (portside)
19 Downhaul (portside)
20 Sheet

Fore topmast staysail

21 Halyard (starboard)
22 Downhaul (starboard)
23 Sheet

Fore course

24 Truss
25 Clew garnet
26 Buntline
27 Leech line
28 Slab line
29 Bowline
30 Lift
31 Brace
32 Sheet

33 Tack
34 Mast tackle (1st)
35 Mast tackle (2nd)
36 Yard tackle tricing line

Fore topsail

37 Tye
38 Buntline
39 Clew line
40 Bowline
41 Sheet
42 Lift & topgallant sheet
43 Reef tackle
44 Brace
45 Shifting backstay

Fore topgallant sail

46 Tye
47 Clew line
48 Buntline
49 Bowline
50 Brace
51 Lift

Fore royal sail

52 Halyard
53 Clew line
54 Lift
55 Brace

Main topmast staysail

56 Halyard
57 Downhaul
58 Tack, fastened to lower end of topmast preventer stay
59 Sheet

Middle staysail

60 Halyard
61 Downhaul
62 Tack, fastened to lower end of jack stay
63 Sheet
64 Tricing line

Main topgallant staysail

65 Halyard
66 Downhaul
67 Tack
68 Sheet

Main royal staysail

69 Halyard
70 Downhaul
71 Tack
72 Sheet

Main course

73 Truss
74 Clew garnet
75 Buntline
76 Leech line
77 Slab line
78 Bowline
79 Lift
80 Brace
81 Sheet
82 Tack
83 Mast tackle
84 Yard tackle tricing line

Main topsail

85 Tye
86 Buntline
87 Clew line
88 Bowline
89 Sheet
90 Lift and topgallant sheet
91 Shifting backstay
92 Reef tackle
93 Brace

Main topgallant sail

94 Tye
95 Clew line
96 Buntline
97 Bowline
98 Brace
99 Lift

Main royal sail

100 Halyard
101 Clew line
102 Lift
103 Brace

Mizzen staysail

104 Halyard
105 Downhaul
106 Brails

107 Tack, fastened to lower end of mizzen preventer stay
108 Sheet

Mizzen topmast staysail

109 Halyard
110 Downhaul
111 Tack
112 Sheet

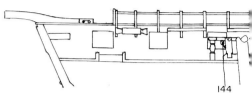

K6

Mizzen topgallant staysail

113 Halyard
114 Downhaul
115 Tricing line
116 Tack
117 Sheet

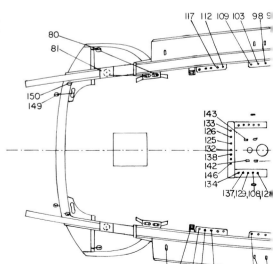

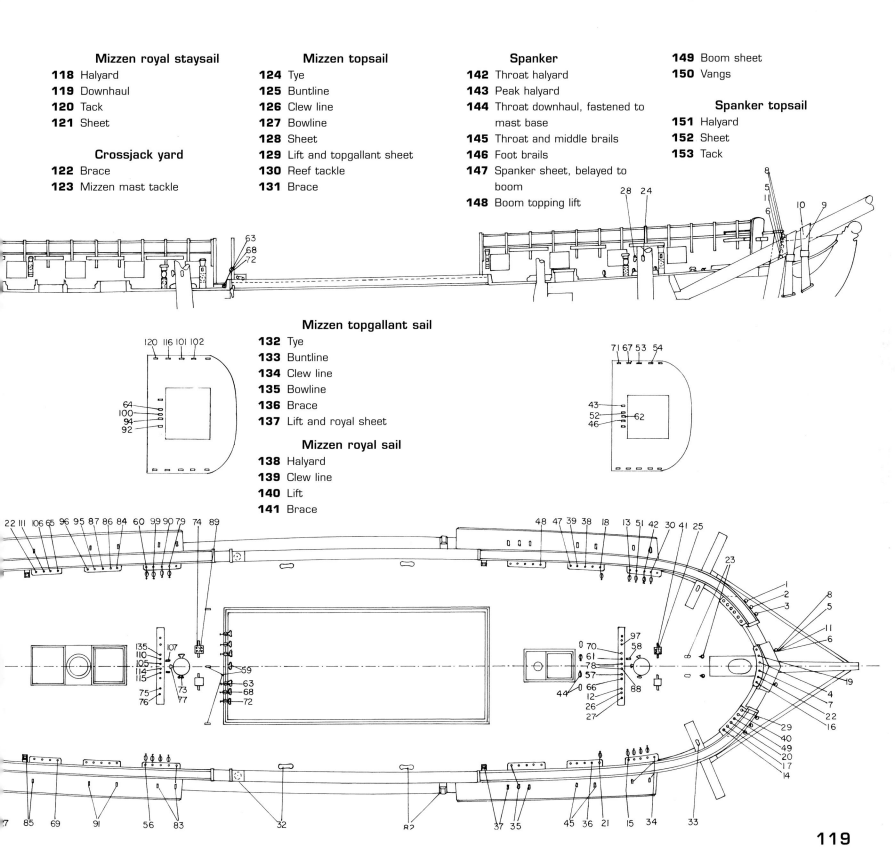

Mizzen royal staysail

118 Halyard
119 Downhaul
120 Tack
121 Sheet

Crossjack yard

122 Brace
123 Mizzen mast tackle

Mizzen topsail

124 Tye
125 Buntline
126 Clew line
127 Bowline
128 Sheet
129 Lift and topgallant sheet
130 Reef tackle
131 Brace

Spanker

142 Throat halyard
143 Peak halyard
144 Throat downhaul, fastened to mast base
145 Throat and middle brails
146 Foot brails
147 Spanker sheet, belayed to boom
148 Boom topping lift

149 Boom sheet
150 Vangs

Spanker topsail

151 Halyard
152 Sheet
153 Tack

Mizzen topgallant sail

132 Tye
133 Buntline
134 Clew line
135 Bowline
136 Brace
137 Lift and royal sheet

Mizzen royal sail

138 Halyard
139 Clew line
140 Lift
141 Brace

119

L Sails

L1 SAIL PLAN AS RIGGED BETWEEN 1803 AND 1815

1 Flying outer jib
2 Flying jib
3 Jib
4 Fore topmast staysail
5 Dotted line on fore mast marks the sails before 1803
6 Fore course
7 Fore topsail
8 Fore topgallant ail
9 Fore royal sail
10 Fore skysails

11 Main course
12 Main topsail
13 Main topgallant sail
14 Main royal sail
15 Main skysail

16 Mizzen topsail
17 Mizzen topgallant sail
18 Mizzen royal sail
19 Mizzen skysail
20 Spanker
21 Spanker topsail

22 Main topmast staysail
23 Middle staysail
24 Main topgallant staysail
25 Main royal staysail
26 Mizzen staysail
27 Mizzen topmast staysail
28 Mizzen topgallant staysail
29 Mizzen royal staysail
30 Fore lower studdingsail
31 Fore top studdingsail
32 Fore topgallant studdingsail
33 Main lower studdingsail
34 Main top studdingsail
35 Main topgallant studdingsail

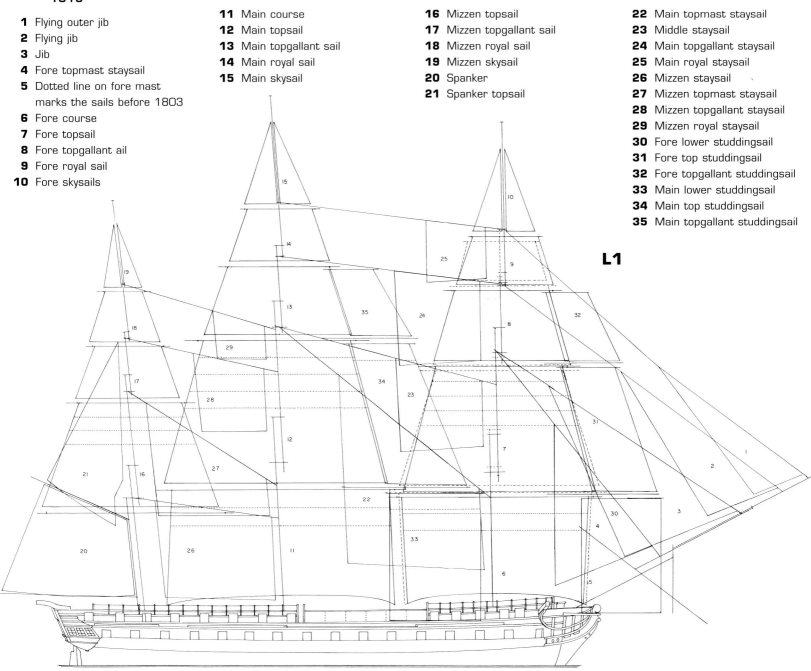

L1

L Sails

L1/1 Head sails

1 Flying jib
2 Outer jib
3 Jib
4 Fore topmast staysail
5 Stay tabling (3–4 1/2in for all)
6 Leech tabling (2–3in for all)
7 Foot tabling (2–2 1/2in for all)
8 Stay boltrope
9 Leech boltrope
10 Foot boltrope
11 Serviced clew boltrope
   (looped and seized)
12 Peak lining
13 Clew lining
14 Tack lining
15 Tabling holes for hanks
   (27in apart)
16 Tack with sewn in thimble
17 Sail cloth (24in wide)
18 Flat seams
   (1 1/4in overlapping)
19 Serviced at clew, tack and
   peak
20 Marling at clew, tack and
   peak
21 Seizing

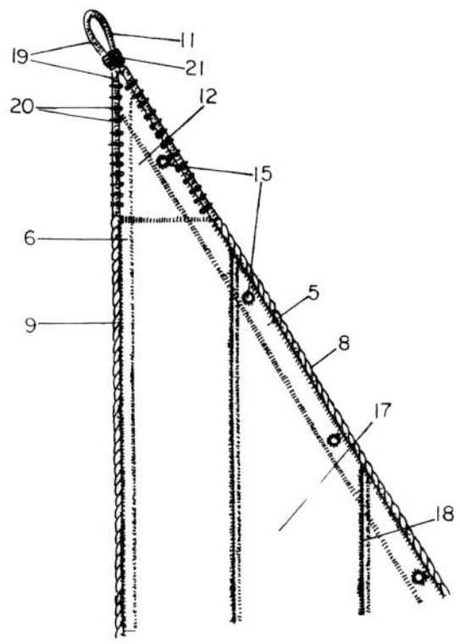

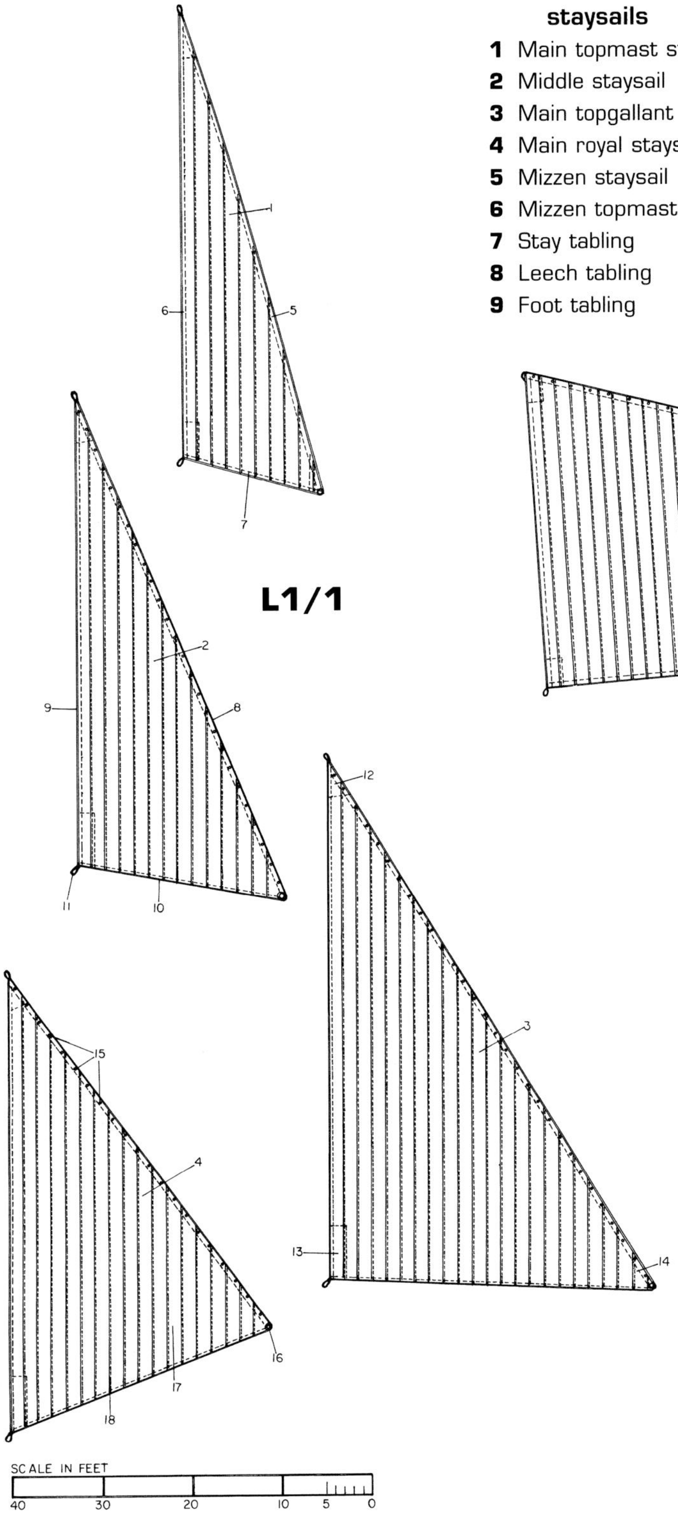

L1/2 Main and mizzen staysails

1 Main topmast staysail
2 Middle staysail
3 Main topgallant staysail
4 Main royal staysail
5 Mizzen staysail
6 Mizzen topmast staysail
7 Stay tabling
8 Leech tabling
9 Foot tabling
10 Mast leech tabling
11 Peak
12 Clew
13 Tack
14 Nock
15 Mast lining
16 Clew lining
17 Peak lining
18 Stay boltrope

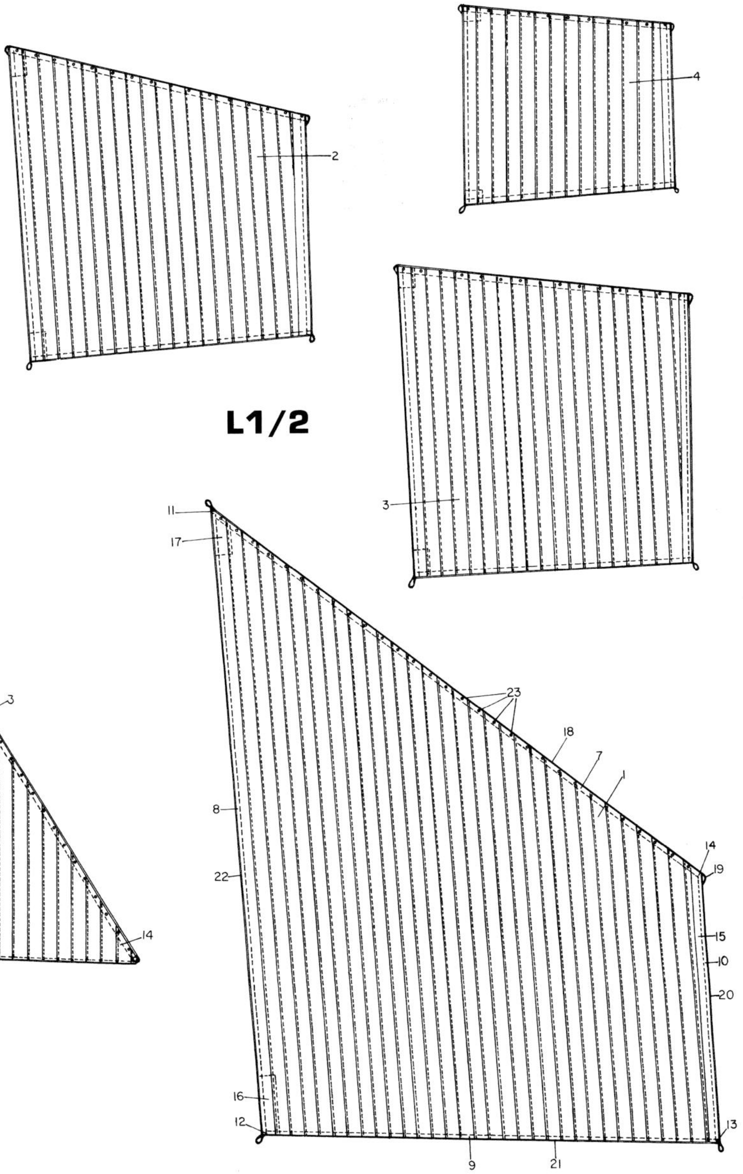

L1/1

L1/2

SCALE IN FEET
40   30   20   10   5   0

L Sails

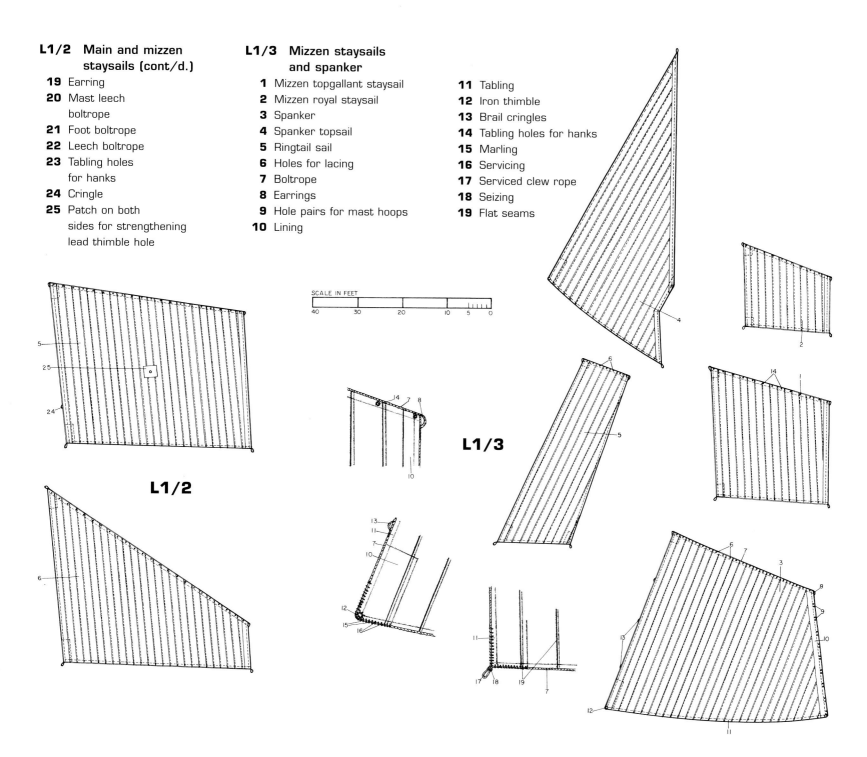

SCALE IN FEET
40 30 20 10 5 0

L1/2

L1/3

L1/4 Fore and main course

1 Fore course
2 Main course
3 Head tabling (4–6in wide)
4 Leech tabling (3–5in wide)
5 Foot tabling (3–5in wide)
6 Side doubling (1 cloth wide)
7 Reef band (1/3 cloth wide)
8 Middle band (2/3–1 cloth wide)
9 Buntline lining (1 cloth wide)
10 Head rope
11 Earrings
12 Leech rope
13 Foot rope
14 Serviced clew rope (loop)
15 Seizing
16 Roband holes in tabling
17 Reef point holes
18 Cringles

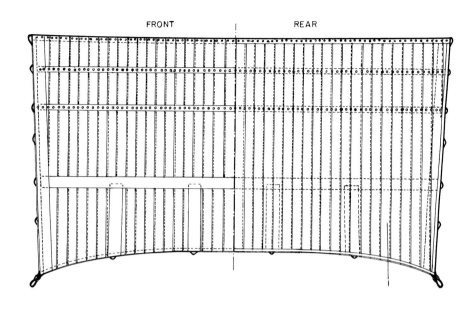

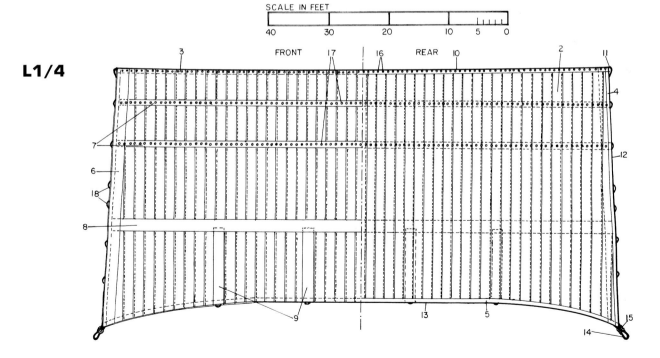

L1/4

SCALE IN FEET

40   30   20   10   5   0

L Sails

L1/5

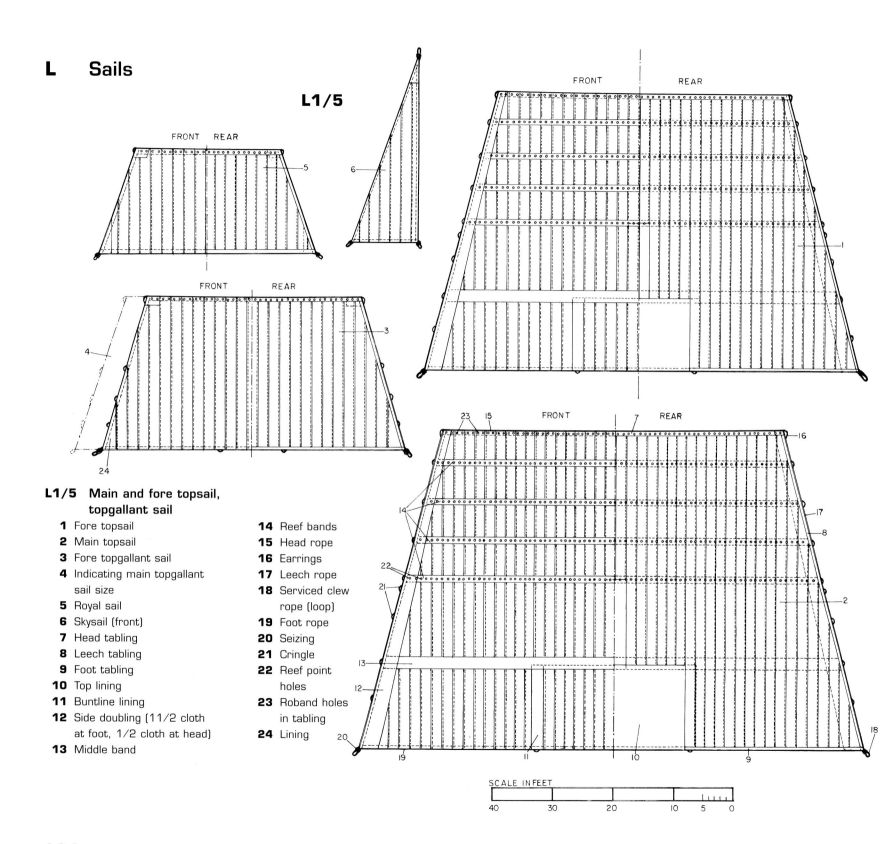

L1/5 Main and fore topsail, topgallant sail

1 Fore topsail
2 Main topsail
3 Fore topgallant sail
4 Indicating main topgallant sail size
5 Royal sail
6 Skysail (front)
7 Head tabling
8 Leech tabling
9 Foot tabling
10 Top lining
11 Buntline lining
12 Side doubling (1 1/2 cloth at foot, 1/2 cloth at head)
13 Middle band

14 Reef bands
15 Head rope
16 Earrings
17 Leech rope
18 Serviced clew rope (loop)
19 Foot rope
20 Seizing
21 Cringle
22 Reef point holes
23 Roband holes in tabling
24 Lining

SCALE IN FEET

40   30   20   10 5 0

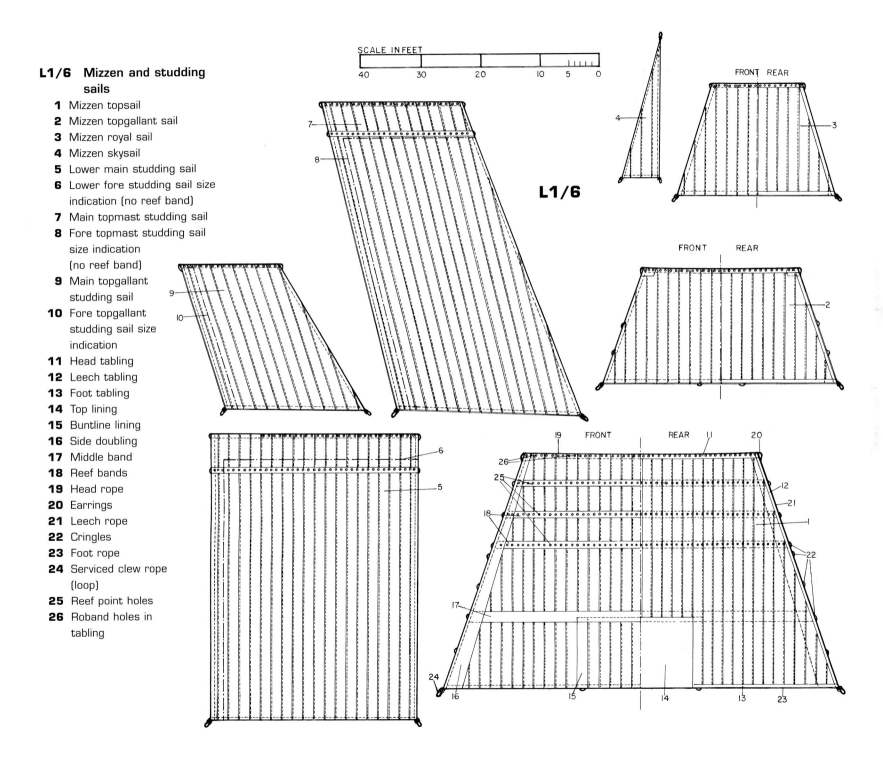

L1/6 Mizzen and studding sails

1 Mizzen topsail
2 Mizzen topgallant sail
3 Mizzen royal sail
4 Mizzen skysail
5 Lower main studding sail
6 Lower fore studding sail size indication (no reef band)
7 Main topmast studding sail
8 Fore topmast studding sail size indication (no reef band)
9 Main topgallant studding sail
10 Fore topgallant studding sail size indication
11 Head tabling
12 Leech tabling
13 Foot tabling
14 Top lining
15 Buntline lining
16 Side doubling
17 Middle band
18 Reef bands
19 Head rope
20 Earrings
21 Leech rope
22 Cringles
23 Foot rope
24 Serviced clew rope (loop)
25 Reef point holes
26 Roband holes in tabling

SCALE IN FEET

40   30   20   10   5   0

L1/6

FRONT REAR

FRONT REAR

FRONT REAR

L Sails

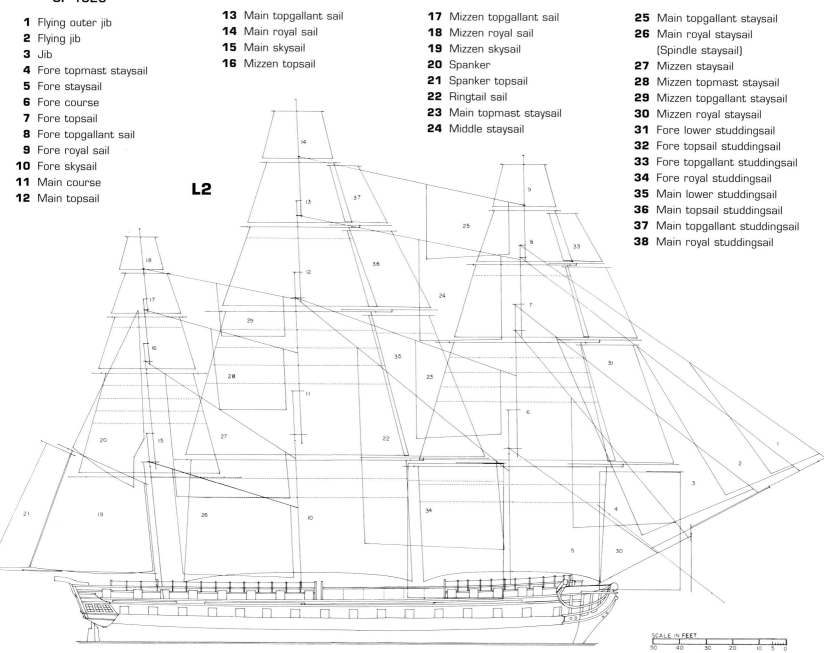

L2

SCALE IN FEET
50 40 30 20 10 5 0

L3   SAIL PLAN
AS RIGGED TO 1997
CONFIGURATION

1 Flying jib
2 Jib
3 Fore topmast staysail
4 Fore course
5 Fore topsail
6 Fore topgallant sail
7 Fore royal sail

8 Main course
9 Main topsail
10 Main topgallant sail
11 Main royal sail
12 Mizzen topsail
13 Mizzen topgallant sail
14 Mizzen royal sail

15 Spanker
16 Spanker topsail
17 Main topmast staysail
18 Main topgallant staysail
19 Main royal (spindlestaysail
20 Mizzen staysail
21 Mizzen topmast staysail

22 Mizzen topgallant staysail
23 Fore lower studdingsail
24 Fore topsail studdingsail
25 Fore topgallant studdingsail
26 Fore royal studdingsail
27 Main topsail studdingsail
28 Main topgallant studdingsail
29 Main royal studdingsail

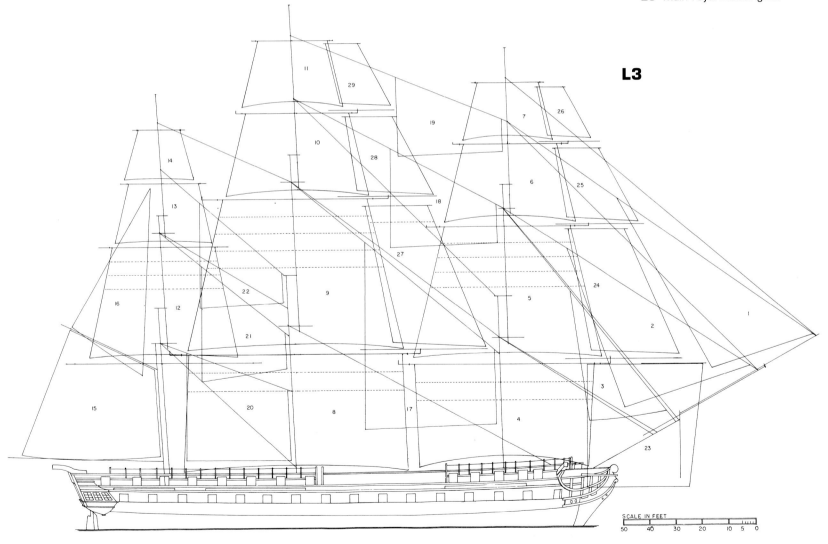

L3

SCALE IN FEET
50   40   30   20   10 5 0

SOURCES

PRIMARY

USS Constitution Engineering Drawings Compact Disk, Naval Historical Center Detachment, Boston, MA

Joshua Humphreys Papers (Notebook), The Historical Society of Pennsylvania, Marked: 'Dreer Hund, September 1927'

PUBLISHED

Bass, W P & E, *Constitution, Super Frigate of many faces, second phase 1802–07* (Shipsresearch Melbourne, Florida, 1981)

Chapelle, H I, *The History of the American Sailing Navy* (Bonanza Books, New York, 1959)

Chapelle, H I, *The History of American Sailing Ships* (Bonanza Books, New York, 1935)

Maffeo, St, Cmdr. ret. USNR, *USS Constitution Timeline* (www.ussconstitution.navy.mil/historyupdat.htm)

Further Reading: see Bibliography

BIBLIOGRAPHY

Brady, W N, *The Kedge Anchor or Young Sailor's Assistant* (New York, 1876; reprint MacDonald and Jane's, London, 1970)

Congreve, W, *An Elementary Treatise on the Mounting of Naval Ordnance* (London, 1811; reprint Frederick Muller Ltd, London, 1970, for Museum Restoration Service Ottawa, Ontario)

Gill, C S, *Steel's Elements of Mastmaking, Sailmaking and Rigging (from the 1794 Edition)* (Edward W Sweetman, New York, 1928)

Goodwin, P, *The Construction and Fitting of the Sailing Man of War 1650–1850* (Conway Maritime Press, London, 1987)

Gruppe, H E, *The Frigates* (Time-Life Books, Amsterdam, 1979)

Heine, W C, *Historic Ships of the World* (Rigby Limited, Adelaide SA, 1977)

Hogg, I & Batchelor, J, *Naval Gun* (Blandford Press, Poole, Dorset, 1978)

Humphreys, J, *A Collection of handwritten Lists and Shipbuilding Instructions* (Dreer Hund, The Historical Society of Pennsylvania, Sept 1927)

Jobé, J, ed., *The Great Age of Sail* (Edita SA, Lausanne, 1967)

King, I H, *The Coastguard under Sail* (Naval Institute Press, Annapolis, 1989)

Lavery, B, *The Arming and Fitting of English Ships of War 1600–1815* (Conway Maritime Press, London, 1987)

Marquardt, K H, *Eighteenth-century Rigs & Rigging* (Conway Maritime Press Limited, London, 1992)

Marsden, L, 'Restoring OLD IRONSIDES' *National Geographic* Vol191 No 6 (Washington DC, 1997)

Naval Historical Centre Detachment, Boston, *Engineering Drawings CD*

Padfield, P, *Guns at Sea* (Hugh Evelyn, London, 1973)

Rees, A, *Rees's Naval Architecture 1819–20* (David & Charles Reprints, Newton Abbot, 1970)

Robertson, F L, *The Evolution of Naval Armament* (Harold T Storey, London, 1968)

Smith, Dr W, *Flags through the Ages and across the World* (McGraw-Hill Book Co UK Ltd, Maidenhead, 1975)

Smyth, Admiral W H, *Sailor's Word Book, A Dictionary of Nautical Termes* (First published in 1867; reprint by Conway Maritime Press, London, 1991)

ACKNOWLEDGEMENTS

Finally a well-deserved thank you to the individuals and institutions that helped with photographs, literature and drawings to make this book what it is:

Jim Hanna, Camberwell, Vic.; William Moss, Boston; David Starr, Mentone, Vic.; US National Archives; US Navy; US Navy Historical Centre Detachment, Boston.

Not to forget my dear wife Sonja, who had to endure the long hours and many months of work that went into this project.

A General arrangement (Scale 1/150)

A1 RECONSTRUCTION OF 1812

A1/2 Half breadth plan

A1/3 Body plan

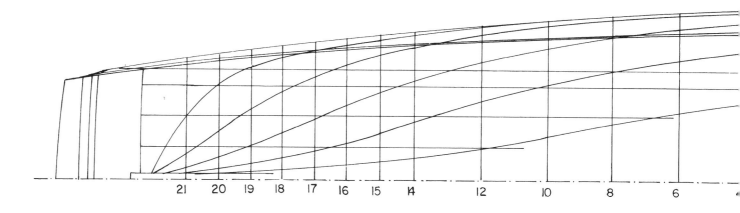

A1/2

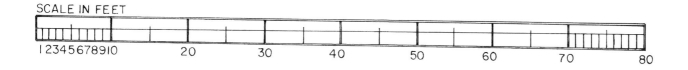

SCALE IN FEET

1 2 3 4 5 6 7 8 9 10   20   30   40   50   60   70   80